Nanotechnology in Cancer

Nanotechnology in Cancer

Edited by

Anshu B. Mathur

The University of Texas MD Anderson Cancer Center, Houston, TX, United States

ELSEVIER

AMSTERDAM • BOSTON • HEIDELBERG • LONDON
NEW YORK • OXFORD • PARIS • SAN DIEGO
SAN FRANCISCO • SINGAPORE • SYDNEY • TOKYO

Elsevier
Radarweg 29, PO Box 211, 1000 AE Amsterdam, Netherlands
The Boulevard, Langford Lane, Kidlington, Oxford, OX5 1GB, United Kingdom
50 Hampshire Street, 5th Floor, Cambridge, MA 02139, United States

British Library Cataloguing-in-Publication Data
A catalogue record for this book is available from the British Library

Library of Congress Cataloging-in-Publication Data
A catalog record for this book is available from the Library of Congress

ISBN: 978-0-323-39080-4

For Information on all Elsevier publications
visit our website at https://www.elsevier.com

Publisher: Matthew Deans
Acquisition Editor: Simon Holt
Editorial Project Manager: Sabrina Webber
Production Project Manager: Nicky Carter
Designer: Greg Harris

Typeset by MPS Limited, Chennai, India

Contents

Contributors

C.L. Altan
Yeditepe University, Istanbul, Turkey

B. Blumenauer
The University of Texas MD Anderson Cancer Center, Houston, TX, United States

S. Bucak
Yeditepe University, Istanbul, Turkey

C.E. Butler
The University of Texas MD Anderson Cancer Center, Houston, TX, United States

M. Caffo
University of Messina, Messina, Italy

G. Caruso
University of Messina, Messina, Italy

M.W. Clemens
The University of Texas MD Anderson Cancer Center, Houston, TX, United States

S.J. Corr
Bayor College of Medicine, Houston, TX, United States; Rice University, Houston, TX, United States; University of Houston, Houston, TX, United States

S.L. Craig
Medical University of South Carolina, Charleston, SC, United States

S.A. Curley
Bayor College of Medicine, Houston, TX, United States; Rice University, Houston, TX, United States

S.F. De Lacey
Kids Development Foundation, London, United Kingdom; University of Zurich, Zurich, Switzerland

L.W. Dunne
The University of Texas MD Anderson Cancer Center, Houston, TX, United States

J.R. Hubenak
The University of Texas MD Anderson Cancer Center, Houston, TX, United States

T.S. Iyyanki
The University of Texas MD Anderson Cancer Center, Houston, TX, United States

V.B. Jensen
The University of Texas MD Anderson Cancer Center, Houston, TX, United States

R. Jeyaraj
University College London (UCL), London, United Kingdom

A.B. Mathur
The University of Texas MD Anderson Cancer Center, Houston, TX, United States

L. Merlo
University of Messina, Messina, Italy

D. Mishra
The University of Texas MD Anderson Cancer Center, Houston, TX, United States

D. Srinivasa
University of Michigan, Ann Arbor, MI, United States

A. Tan
University College London (UCL), London, United Kingdom; Stanford University, Stanford, CA, United States; Kids Development Foundation, London, United Kingdom

Q. Zhang
The University of Texas MD Anderson Cancer Center, Houston, TX, United States

About the Authors

Cem L. Altan, PhD

Cem Levent Altan received his PhD in Chemical Engineering from Yeditepe University (Turkey) and Eindhoven University of Technology (Netherlands). Currently, he is an Assistant Professor at the Department of Chemical Engineering at Yeditepe University in Istanbul. His research focuses on synthesis and control over the formation of magnetite nanoparticles, magnetic hyperthermia, and thermophysical properties of magnetic nanofluids.

Brian Blumenauer

Brian Blumenauer is a third-year general surgery resident at the University of Texas Medical School in Houston. Upon completion of his general surgery residency, Dr. Blumenauer will look to obtain a fellowship position in the field of plastic and reconstructive surgery.

Mark W. Clemens, MD, FACS

Mark W. Clemens is in the Plastic Surgery faculty of MD Anderson Cancer Center, University of Texas in Houston. He completed plastic surgical training at Georgetown University, research training programs at the National Institutes of Health (NIH) and Harvard Children's Hospital, and received a microvascular reconstructive surgery fellowship at The University of Texas MD Anderson Cancer Center.

He has lectured internationally and written more than 50 peer-reviewed articles and 30 book chapters. He serves on study section of the Plastic Surgery Foundation (PSF), the editorial board of the *Aesthetic Surgery Journal*, and is a board member for the Aesthetic Surgery Education Research Foundation (ASERF). He was a member of the surgical team that performed the world's first skull and scalp transplantation. His research interests include robotic, microvascular, and prosthetic reconstructive surgery of the breast. He is the MD Anderson site principal investigator for the NIH R01-funded Mastectomy Reconstruction Outcomes Consortium (MROC) Study. He leads a multidisciplinary research team and tissue repository focused on the study of breast implant-associated anaplastic large-cell lymphoma (BI-ALCL). He serves as an American Society of Plastic Surgeons (ASPS) society liaison to the US Food and Drug Administration, and chairs a subcommittee for ASPS overseeing national research and education efforts for BI-ALCL.

Charles E. Butler, MD, FACS

Dr. Charles E. Butler is Chairman and Professor in the Department of Plastic Surgery at The University of Texas MD Anderson Cancer Center. Dr. Butler is board certified in both general and plastic surgery. He is a well-recognized expert in reconstructive surgery, particularly in complex reconstruction and microsurgery with over 300 publications and 700 invited national and international presentations, instructional courses, and lectures. He has clinical and research interest and expertise in torso reconstruction, including the abdominal wall, breast, chest wall, pelvis, and spine. He is an active clinical and basic science researcher and mentor. Dr. Butler is an international expert in complex torso reconstruction, microsurgery, flap, component separation, implantable biologic matrices, and tissue engineered constructs.

Seyda Bucak

Seyda Bucak received her PhD in Chemistry from University of East Anglia, UK, started working on magnetic nanoparticles during her postdoctoral research at MIT. Since then, her work is predominantly on the synthesis, characterization, and applications of magnetic nanoparticles in biomedical field and material science. She has been working at the Department of Chemical Engineering at Yeditepe University since 2004, where is currently a full-time Professor and head of department.

Maria Caffo, MD

Dr. Maria Caffo (University of Messina, School Medicine, 1994) is currently Assistant Professor in Neurosurgery at Department of Neurosurgery, University of Messina, Italy. She has been a Member of the Italian Society of Neurosurgery since 2000, a Member of the Italian Association of Neurooncology since 2000, and a Member of the European Association of Neurooncology since 2002. Dr. Caffo has a PhD in Neurooncology. She is also Professor of Physiology in the degree course of Technique of Neurophysiopathology and in the degree course of Science in Nursing, and Professor of Neurooncology in the Program Residency in Neurosurgery, Department of Neuroscience, University of Messina, Italy, and she is a reviewer for several technical journals.

Gerardo Caruso

Gerardo Caruso (University of Messina, School Medicine, 1990) is currently an Assistant Neurosurgeon, Neurosurgical Clinic, University of Messina, Italy. He has been a Member of the Italian Society of Neurosurgery since 1996 and a Member of the American Nanosociety since 2011. Dr. Caruso previously served as an officer in the Italian Navy and has also served as a reviewer for a number of international technical journals. He is author of various papers refereed on international scientific journals.

Stuart J. Corr

Dr. Stuart J. Corr is an Assistant Professor of surgery in the Michael E. Debakey Department of Surgery at Baylor College of Medicine. He also holds an adjunct Assistant Professor title at Department of Bioengineering, University of Houston. He has a background in electronic-, system-, and nano-engineering, as well as nano-mediated noninvasive radio frequency cancer hyperthermia. He is currently involved in the design and prototyping of 3D printer technologies and is Director for the Baylor College of Medicine, Department of Surgery Incubator for Surgical Innovation.

Suzanne L. Craig, DVM, DACLAM, MBA
Suzanne Craig is currently Department Chair of Comparative Medicine and Director of Laboratory Animal Resources at the Medical University of South Carolina. She has published several peer-reviewed articles and book chapters including work on nanoparticles and cancer. She is the Past-President of the American Society of Laboratory Animal Practitioners. She formerly was Deputy Department Chair at The University of Texas MD Anderson Cancer Center of the Department of Veterinary Medicine and Surgery. She has held roles as Director of the Gulf Coast Consortium for the training of veterinary residents in comparative medicine for which she received an NIH grant, Director of Education, Chief IACUC Veterinarian, Section Chief, and other positions. She has been an invited speaker both nationally and internationally.

Steven A. Curley, MD, FACS
Dr. Steven A. Curley is Professor of Surgery and Chief of Surgical Oncology at Baylor College of Medicine. He is also an adjunct Professor in Department of Mechanical Engineering and Materials Science at Rice University. He leads a translational research laboratory at Baylor College of Medicine and has been involved with the development and design of several FDA-approved medical devices. His work currently focuses on the clinical translation of noninvasive radio frequency cancer hyperthermia therapy.

Sherrie F. De Lacey
Sherrie F. De Lacey is a doctoral student, currently studying PhD in International Business at the University of Zurich, Switzerland. She previously worked in the City of London as an analyst, developing proprietary trading algorithms, helping to raise money and broker deals. She has now founded Ranait Fleur and often volunteers with children; being very well traveled having lived on four continents mostly led by her passion for safeguarding children, Sherrie is CFO at Kids Development Foundation.

Lina W. Dunne
Lina W. Dunne is based at the University of Texas MD Anderson Cancer Center, Department of Plastic Surgery, Houston, TX, United States.

Justin R. Hubenak
Justin R. Hubenak is based at the University of Texas MD Anderson Cancer Center, Houston, TX, United States.

Tejaswi Sriranga Iyyanki

Tejaswi Sriranga Iyyanki is a graduate research assistant and a third-year PhD–MBA student at Penn State College of medicine in the biomedical sciences program. His current research interests include genomics, epigenomics, cancer, stem cells, big-data biology, and next-generation sequencing technologies. He previously worked at the University of Texas MD Anderson Cancer Center as a research assistant in plastic surgery.

V. Behrana Jensen, DVM, DACLAM

Vanessa Jensen is an Associate Professor and Section Chief in the Department of Veterinary Medicine and Surgery at The University of Texas MD Anderson Cancer Center. She is a Diplomate of the American College of Laboratory Animal Medicine. She received her doctorate in Veterinary Medicine from Iowa State University and was a Post-Doctoral Research Associate in the Department of Immunology and Microbiology at The University of Iowa. She has been involved with publications in various fields of research involving laboratory animals.

Rebecca Jeyaraj, MBBS

Rebecca Jeyeraj is an MBBS medical student at University College London (UCL), currently in her clinical years. She obtained her integrated BSc in Clinical Sciences from UCL, graduating with First Class Honours, and was also awarded the Faculty of Medical Sciences Medal. Her research interests center around the application of nanoscience to clinical and surgical medicine.

Anshu B. Mathur, PhD

Anshu B. Mathur holds a position as a tenured Associate Professor in the Department of Plastic Surgery at The University of Texas MD Anderson Cancer Center, Houston, Texas, United States. She also serves as the Director of Research in the Department of Plastic Surgery for the Tissue Regeneration and Molecular Cell Engineering Labs (TRAMCEL). The TRAMCEL provide a unique interdisciplinary clinically translatable research environment, opportunities, and platform for training of engineering and surgery research trainees to go from nanoscale to clinic. We have cultivated expertise in micro/nano technologies and devices as they are applied to the fabrication of biomaterials and therapeutic delivery vehicles with applications in the fields of nanomedicine and regenerative medicine.

She received her doctorate degree in Biomedical Engineering from Duke University, Durham, North Carolina, United States, before joining the MD Anderson Cancer Center faculty as a tenure-track Assistant Professor in 2003. She also holds two Bachelor of Science degrees with honors and one Master of Science degree from North Carolina State University, Raleigh, North Carolina, United States, and another Master of Science degree from Duke University, Durham, North Carolina, United States.

Lucia Merlo

Lucia Merlo (University of Messina, School of Medicine, 2001) is a Neurosurgeon. She has been a Member of the European Association of Neurosurgical Societies since 2010. She is interested in Neurooncology, Pediatrics, and Functional Neurosurgery.

Deepa Mishra

Deepa Mishra is based at the University of Texas MD Anderson Cancer Center, Houston, TX, United States.

Dhivya Srinivasa

Diplomate, American Board of Surgery. Surgical Internship, University of California, San Francisco 2010–11, General Surgery Residency at University of Texas, Houston 2011–15, and Plastic Surgery Resident, University of Michigan 2015 to current.

Aaron Tan, MBBS

Aaron Tan is an MBBS medical student at University College London (UCL), currently in his clinical years. He is concurrently a visiting scholar at the Biomaterials and Advanced Drug Delivery Laboratory (BioADD) at Stanford University. He holds two degrees from UCL, a BSc (Hons) in Surgical Sciences and a PhD in Surgery and Bioengineering. Aaron has authored in excess of 40 PubMed-indexed publications, with a total citation of 1051, and a h-index of 16.

Qixu Zhang, MD, Ph.D

Qixu Zhang is an Assistant Professor in the Department of Plastic Surgery at University of Texas MD Anderson Cancer Center. He received medical doctorate degree and completed plastic surgery residency training at Peking Union Medical College in China. Dr. Zhang completed his postdoctoral fellowship at UTMB and MDACC in Houston. He is an active member of PSRC, WHS, SFB, TERMIS, and SPRBM. His research focuses on the large-scale composite soft tissue engineering using natural platform and adult stem cells for reconstructive surgery. His work has been published in high impact journals in the field including *Biomaterials, Acta biomaterialia, JID*.

Foreword

Application of nanotechnology in cancer is the forefront of nanomedicine that can improve quality of life of the patients suffering from disease, disability, and disfigurement. Nanotechnology can be applied to target diverse areas afflicted by cancer such as invasive and metastatic diseases, aesthetic plastic surgery, and therapeutic delivery via catheters and stent coatings.

Nanotechnology can generate various structures and shapes with programmable properties resulting in specificity of function. While research has heavily focused on carbon and gold nanoparticles, nanotubes, and other combination structures, the size of the structures defines it as nano in the range of 1–100 nm as per definition of the National Institutes of Health (https://www.nih.gov/research-training/nanotechnology-nih), Bethesda, Maryland, USA.

This edition of the book written from a major cancer center, The University of Texas MD Anderson Cancer Center, Houston, Texas, USA, has attempted to cover the entire range of ongoing work in Nanotechnology. A major area that my group has written about in the past is Imaging Nanotechnology and that is not covered at length in this book.

Cancer is treated as compromised wound bed with issues such as heterogeneous vasculature, genetic malformations, cellular diversity, microbe infiltration, and other issues that affect its growth and preservation in the human body. Although cancer bed is a challenging environment to begin with the invasive and metastatic diseases make it impossible to treat.

The targeting of the nanotechnology-based therapeutics is multifaceted and can be customized with the specificity of the therapeutic. Further customization by surface modification techniques allows receptor–ligand specificities toward a tumor bed. Free energy of molecules in that regime is affected by entropy, disorder. The targeted therapeutic area is loaded with biological sources and synthetic molecules. Specifically, therapeutic industry such as the nutraceuticals has roots in Ayurveda medicine. Combination targeting with factors considered from ancient Ayurveda and free energy targeting by surface modifications has implications in cancer treatment.

The existence of cancer has deep roots in medicine and the complexity of the tumor and treatment may be tied together. Similar to gene therapy in the past, nanotechnology is considered a forefront of medicine targeting molecular regimes that are unexplored but challenging. Considering the factors affecting tumor treatment using nanotechnology, the science and its translation is controversial similar to gene therapy. The failure rates could be considerable if the complex molecular nature is not identified.

Anshu B. Mathur
The University of Texas MD Anderson Cancer Center,
Houston, TX, United States

Preface

I dedicate this book to my children, Roshan, Aarushi, Devan, and the one who didn't make it.

I have been working in the area of Nanotechnology applications in cancer since my appointment at The University of Texas MD Anderson Cancer Center as a tenure-track Assistant Professor in 2003. While the cancer center is rich in knowledge about cancers, I was able to collaborate with therapeutic researchers and offer my expertise as a Biomedical Engineer and build molecules such as silk fibroin–coated liposomal emodin and nanocurcumin. We have been conducting studies to assess absorption of the nanocurcumin in an in vivo rat tumor model and will continue collaborations with MD Anderson clinical and basic science collaborators. The advantages of this nanoparticle approach for clinical health applications are tremendous, such as ease of manufacturing, versatility toward any therapeutic, high efficacy against diseased cells, biodegradability, no toxic byproducts, and biodisposability.

The first edition of this book focuses on nanotechnology in cancer and has further impact in regenerative medicine with respect to complete cancer care. Combination of nanotechnology with cell and tissue engineering is the forefront in regenerative nanomedicine that can help understand disease at the nanoscale leading to novel therapy modalities and improve quality of life in cancer patients. I have cultivated multiphase strategies to address the restoration of functional tissues in cancer patients with critical size defects, by focusing my research on the development of nanoengineered regenerative biomaterials and targeted nanotherapeutics for regulated local repair. The ultimate goal of my research group is to engineer regenerative, cell-responsive, and therapeutic three-dimensional nanostructured biomaterial matrices that would recruit, guide, and renew cells at the repair site by surrendering control of the regenerative process to the cells, which regulate the process at the cell–matrix nanointerface (<100 nm). In order for investigators in this emerging field of nanomedicine and regenerative medicine to develop complete repair strategies with high efficacy, we will first have to understand the rules at the nanoscale as they pertain to tumors and regenerative stem cells. What are the nanosignatures or the quantum molecular signatures of various cells? The quantum molecular signatures of the tumor cells that are intermixed with stem cells at a repair site can help us differentiate the "target sites" on tumor cells versus the regenerating stem cells at the same location. With the assistance of microsurgery, we can reach these "targets" in order to detect, manipulate, and guide local regeneration.

The future of nanotechnology at MD Anderson Cancer Center lies in continuing to develop collaborative efforts across the institution in order to study the regulation of cells at a molecular scale to achieve repair, reconstruction, and regeneration in cancer patients via molecularly engineered materials. I plan to lead initiatives at the national level to build this area of research and develop a translational device platform in the area of regenerative materials and quantum-targeted nanotherapeutics.

I would like to thank my Sr. Administrative Assistant, Ms. Micquelyn Titus, and the Department of Plastic Surgery for their administrative support. The Elsevier team and the publisher have provided publishing infrastructure for me to be able to edit this book. I would also like to acknowledge my family.

This book brings together my work since 2003 at UT MD Anderson Cancer Center and I hope that the community can grasp how the two independent areas of Nanotechnology and Cancer come together.

Anshu B. Mathur
The University of Texas MD Anderson Cancer Center,
Houston, TX, United States

Introduction to Bio-Nanotechnology

The properties of cells and tissues, whether they are diseased like cancer or healthy, are a function of their interactions with each other and have to be treated as an integrated closed system. The key is to recognize that the macrotissue properties are driven by its microenvironment, which is affected at the nanoscale by an integrated cellular assembly and its mechanical interactions with the other cells and matrix components. From an engineer's perspective, the divide in knowledge between cancer and health is due to the divide between engineers/biophysicists and cancer cell biologists working independently in their own research microculture. For example, engineering methods of study in these cancer systems could get trapped in in vitro 2D monolayer experiments and new systems designed by engineers/biophysicists who probe cells at nanoscale and below with instrumentations such as surface force apparatus and atomic force microscopy or model such effects using finite element modeling at the microscale may not translate into the cancer world. The fanciest experiments are in 3D in situ live animal imaging with multiphoton second harmonic generation confocal imaging, which gives structural information and cell–cell interactions at a global tissue level. Cancer cell biologist can break the cellular pathways down to their microRNA but compromise the cohesiveness of the integrated cellular unit. The next generation of instrumentations have to be developed in collaboration between cancer cell biologist, who have the molecular knowledge of which molecules are being expressed by the cells, and engineers/biophysicists, who know how to combine the latest technologies that will allow probing of cells at the nanoscales and below in their native 3D microenvironments, such that differences between cancer and normal tissues properties can be built into a library that has quantitative information about an integrated cell–matrix assembly. For example, tumor type (breast, liver, pancreas, etc.), cell type (epithelial cell, hepatocyte, endothelial cell, etc.), Ligand X interacts with Receptor X on the cell with an adhesion strength of Y and drives actin microfilament cytoskeletal assembly to form stress fibers that may increase the stiffness *Z* of the cell globally, as the signal travels down to the cell–integrin interactions that leads to integrin clustering, which feeds back into the cytoskeletal stress fiber assembly and perhaps increases/decreases/balances the expression of Receptor X. If we have all of this information about integrated closed tumor systems, we can develop targeted nanotherapeutics with minimal toxicity and advance our knowledge for treatment of other diseases also.

Engineers have developed imaging technologies such as quantum dots (QDs) to overcome issues with cancer microenvironment. Although QDs have unique optical properties, such as size tunable absorption and emission at various wavelengths, improved photostability, and narrow emission peak to enable multiplexing, the toxicity associated with QDs in biological systems is a major shortcoming that can be overcome using biological coatings. For example, the synthesis, characterization, and application of silk fibroin (SF)-coated semiconductor nanocrystals, a.k.a. QDs, in cellular systems was reported. The coating of QDs with SF provides a biological and biocompatible alternative to traditional fluorescent markers for in vitro and in vivo cellular imaging applications. The biocompatible interface provided by the SF coatings would be expected to assist in the clearance of SF-coated QDs. These features increase the potential for administration of SF-coated QDs for in vivo imaging administration for versatile disease-specific imaging needs. From a nanoparticle size perspective, it is interesting to note that SF has the molecular composition and structural features to allow it to coat particles <10 and >100 nm.

Biological nanocoatings have also been used to overcome numerous barriers to drug delivery into the tumor site. Careful considerations are required when designing a new "capsule" formulation of a drug for high efficacy at the tumor site. The success of a therapeutic/drug is dependent upon its mode of delivery and its potency at the site of tumors. At the site of a solid tumor, key issues that make the drug delivery to tumors a challenging affair are heterogeneous vascular architecture and permeability, high interstitial pressures in the necrotic core, large interstitial distance between the tissue mass and vessels, low convective transport, acidic pH, hypoxia, and lack of lymphatic drainage. In addition, due to lack of specificity of the delivery device for tumor cells, administration of high dosage of the drug causes drug resistance and toxicity to normal tissues. Thus biological nanocoatings have been developed that enhance targeting to a tumor cells, increase retention of the drug, release drug in a controlled manner, and enhance the efficacy of a drug/therapeutic. The integration of therapeutic that may have targeted specificity for a tumor type with a delivery device that does not compromise its specificity for the tumor is a critical parameter to consider.

Treatment modality to revolutionize the way cancer is treated currently is to stay with a comprehensive approach yet maximize targeting and specificity. The entire approach and its individual components are innovative in their own respect.

CHAPTER

Gold nanoparticles for noninvasive radiofrequency cancer hyperthermia

1

S.J. Corr[1,2,3] and S.A. Curley[1,2]

[1]Bayor College of Medicine, Houston, TX, United States [2]Rice University, Houston, TX, United States [3]University of Houston, Houston, TX, United States

CHAPTER OUTLINE

INTRODUCTION

The use of gold nanoparticles (AuNPs) to deliver targeted, controlled hyperthermia is currently under investigation in the context of noninvasive radiofrequency (RF) cancer hyperthermia. It is thought that once these AuNPs have internalized within cancer cells, the exposure of the cells to an external, noninvasive (and non-ionizing) RF electric field will initiate hyperthermia within those cancer cells only, leaving the neighboring healthy cells unaffected. A simplified overview of this process can be seen in Fig. 1.1.

AuNPs have generally been the material of choice for this application over the last decade due to their well-established, facile synthesis protocols, which allow for a variety of shapes and sizes to be fashioned, as well as their bioinert nature,

Nanotechnology in Cancer.

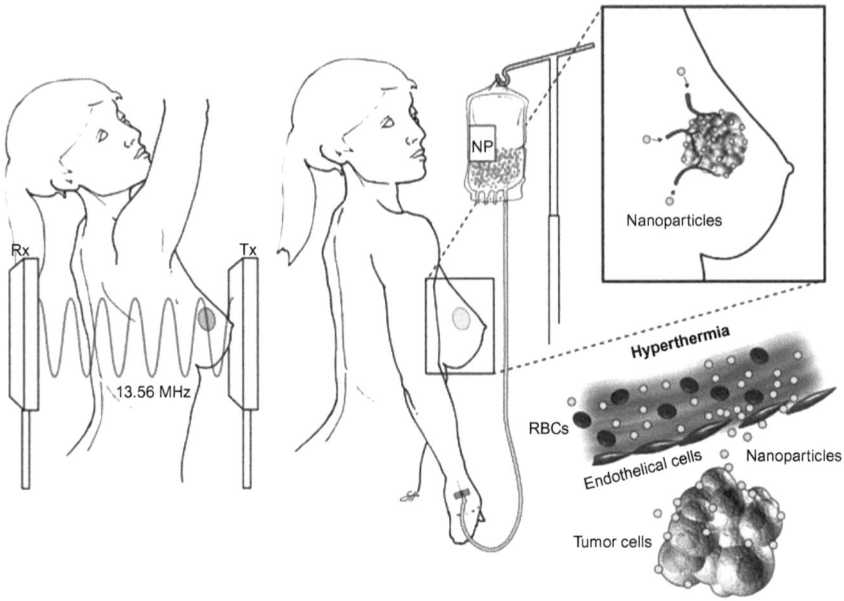

FIGURE 1.1

Noninvasive RF system for cancer therapy. The patient is placed between the transmitting (Tx) and receiving (Rx) heads where a 13.56-MHz high-power electric field is generated. A breast tumor is shown in yellow (gray ellipsoid in print versions). As the tumor gets heated due to absorption of RF waves, the blood vessels dilate and become more permeable, increasing tumor blood flow and extravasation of macromolecules, chemotherapeutics, or nanoparticles from the vessels into the tumor tissue where they are internalized by cancer cells.

although this is still under debate. Due to their interesting electrical and conductivity properties, they were also deemed a perfect starting material as their interactions with electric fields could potentially elicit heating due to Joule heating mechanisms. As we will show however, this was not necessarily the case and indeed there has been intense speculation as to whether AuNPs are "hot or not" in the presence of an RF electric field.

Although the exact heating mechanisms of AuNPs are still under debate and are primarily studied in simplified aqueous solutions, there is an increasing amount of in vitro and in vivo evidence to suggest that AuNPs, especially those that are conjugated to antibodies and cancer chemotherapeutics such as C225 (Cetuximab) and gemcitabine (GCB), still have a place in the field of noninvasive RF cancer hyperthermia, and warrants further investigation.

BRIEF OVERVIEW OF CLINICALLY AVAILABLE HYPERTHERMIA SYSTEMS

The area of electric field cancer hyperthermia has been under investigation over the last century as a clinically relevant means of inducing noninvasive hyperthermia in patients with a variety of cancer types. A comprehensive review of the field can be found in the literature [1]. The treatment is based on the heat-inducing interactions of multifrequency electric field energy with biological tissues based on the nonzero dielectric properties of tissues, organs, and cancerous lesions (i.e., electrical energy is absorbed by the tissues and converted to heat, inducing hyperthermia).

Heating characteristics are dependent on the frequency of the electric field with most heating interactions occurring between 1 MHz and 1 GHz. As the frequency of the electric field increases, the penetration depths of the electric fields into the patient decrease. Hence, lower MHz frequencies are commonly used for the treatment of deep-seated tumors, while higher frequencies are often utilized for superficial tumor types.

There have been several system types investigated since the 1980s, each with their pros and cons, each with their varying levels of success with treating a variety of cancers. The first "main stream" commercially available clinical RF hyperthermia system was the Thermotron RF-8 (Yamamoto Vinita Co. Ltd., Japan) developed in the late 1970s/early 1980s. This system is an 8-MHz capacitively coupled system that utilizes water-cooled electrodes placed on opposite sides of the patient (i.e., parallel to each other with the intended treatment area "sandwiched" in-between). The body area between the two electrodes acts similar to the dielectric material found in electronic capacitors and is gradually heated as a function of power. Although this system is straightforward and easy to use, it does have inherent disadvantages such as high subcutaneous fat heating, which is alleviated somewhat by the use of a water-cooled "bolus"; as well as the instability of low-frequency RF field and its dependence on electrode size, location, and tissue parameters, which often result in "hot spot" formation. Several studies however have shown the effective use of this system (combined with and without radio- and chemotherapy) in the treatment of a variety of superficial and deep-seated cancer types, providing that the surface cooling and electrode configurations are properly managed [2].

Pyrexar Medical (formerly BSD Medical) has several systems available that include the BSD 500 and the BSD 2000 3D/MR. The BSD 500 is a self-contained treatment system for mobile applications using a fixed 915-MHz signal for treating superficial tumors located approximately 3–6 cm under the skin. On the other hand, the BSD 2000 3D/MR utilizes a 24-dipole antenna-phased array method (operating at 75–140 MHz) to accurately "steer" the electrical energy, and treatment area, within the patient at depths of up to 20 cm. This system can also be integrated into a magnetic resonance imaging (MRI) system to allow for MR thermometry—a means of visualizing the RF-induced temperature increases by analyzing the shift in proton resonance frequency, which is a function of temperature. The ability to measure internal RF-induced temperature changes is of critical importance for successful therapy: hence the MR integration. Clinical studies with this system have also seen some level of success. For example, in a recent randomized phase-3 multicenter study investigating the use of neoadjuvant chemotherapy alone or with regional hyperthermia for localized high-risk soft-tissue sarcoma (STS), it was demonstrated that the addition of hyperthermia increases the benefit of chemotherapy and is a new effective strategy for patients with high-risk STS, including STS with an abdominal or retroperitoneal location [3].

Another example is Celsius42+ GmbH who have developed a 13.56-MHz capacitively coupled system (Celcius TCS Hyperthermia System), which employs the use of two water-cooled electrodes (similar to the Thermotron RF-8 system) that effectively "sandwiched" the treated area between the two electrodes and allow the RF energy to be coupled into the patient. By combining chemotherapy or radiation therapy alongside the RF hyperthermia, they have shown interesting results (randomized phase III studies) in tumors of the head and neck, brain, breast, colon, rectum, and gynecological tumors. For further reference, other companies active in this area include OncoTherm, Andromedic, and Alba Hyperthermia.

The RF system currently under development in our laboratories is shown in Fig. 1.2. The device is based on a cascaded LC network whereby a high-power (100–1500 W) electric field operating at 13.56 MHz is transmitted from the transmitting head into the sample under exposure. The use of a receiving head behind the patient helps focus the RF energy, preventing stray RF. We have also developed a portable, miniaturized version of this system, as can be seen in Fig. 1.3, whereby we can perform simultaneous RF exposure and intravital microscopy, in real-time, to investigate the biological effects of RF on cancer tissues and nanoparticle dynamics. See Ref. [4] for a full overview of this system.

NANOPARTICLE-MEDIATED CANCER HYPERTHERMIA

The size-dependent magnetic, optic, electronic, and biological properties of nanomaterials has been the center of much study since around the year 2000. Many new physical effects emerge when certain material dimensions scale down to the order of nanometers. These include the plasmonic resonance of AuNPs, which can be tuned across the optical spectrum to make the nanoparticles appear in a variety of different colors [5]; the electronic nature of carbon nanotubes (CNTs)—the electronic properties of which can be altered by changing the CNT chirality to be either metallic, semiconducting, or a hybrid of the two [6]; and the

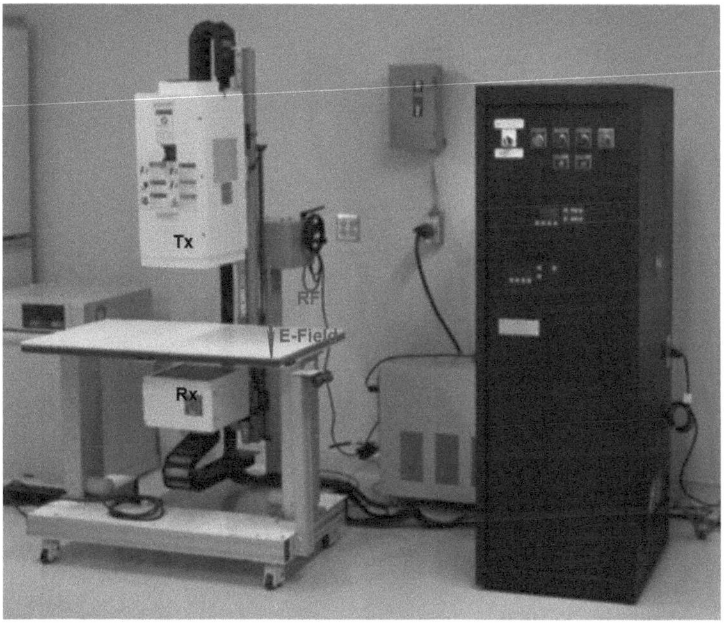

FIGURE 1.2

Noninvasive RF system for cancer therapy (Curley Lab, Department of Surgery, Baylor College of Medicine, Houston, TX). The black chassis (right) houses the RF amplifier and autotune circuitry, which provides power to the impedance matching RF applicator (white) that consists of a transmitting (Tx) and receiving (Rx) head. A high-power (100–1500 W) RF electric field is generated between the Tx and Rx heads. The sample is placed on a Teflon table (to minimize RF interactions) and positioned between the Tx and Rx heads.

enhanced MRI relaxivitiy properties of Gd^{3+} loaded ultrashort CNTs, making these nanoconstructs superlative MRI contrast agents based on nanoscale superparamagnetism and unique water–proton interactions [7,8].

Many of these unique nanoscale physical phenomena are being exploited to allow for new ways of administering hyperthermia-based cancer therapies. These emerging technologies potentially allow for cell-specific, targeted oncology approaches that can either be complimentary or be superlative to current clinical cancer treatments such as chemo-, radio-, immuno-, and proton therapies, and the traditional surgical resection approach. Many approaches so far have been photothermal in nature; i.e., they use the absorption of optical and/or near-infrared light to initiate nanoparticle heating, the energy of which is dissipated through the cancer cell initiating hyperthermia. In these regards, there have been many investigations using AuNPs, CNTs, and graphene.

AuNPs of different sizes, shapes, surface functionalization, and alloy compositions (i.e., SiO_2 core with Au shell) have been of prominent interest in this field [9] due to their facile synthesis techniques and biological inertness (although this

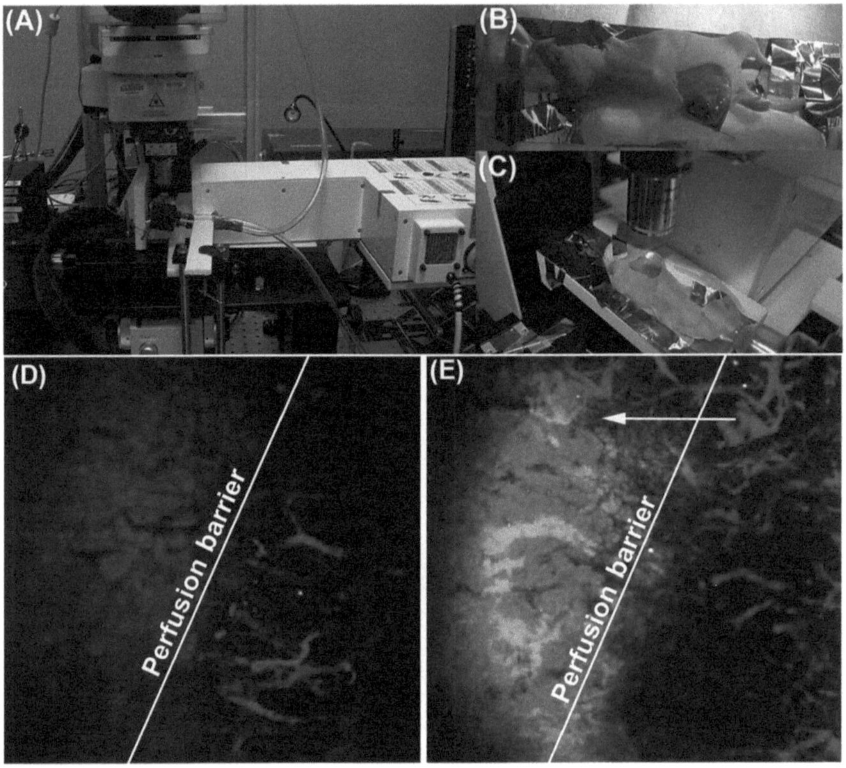

FIGURE 1.3

(A) Portable RF system retrofitted to an intravital microscope to allow for real-time, simultaneous imaging of RF effects on biological tissues and blood perfusion. (B) Mouse manipulation for imaging whereby an incision is made to expose and gently manipulate the tumor for IVM imaging. (C) Tumor under IVM illumination with a 4 × objective lens. (D) RF exposure shows transport of fluorescently bound albumin (blue (light gray in print versions)) across the perfusion barrier into tumor region (red (dark gray in print versions)) after ~4 min of RF exposure at 200 W (E).

is still the subject of much debate). The ability to tune the surface plasmonic resonance (SPR[1]) to directly coincide with the excitation wavelength is an extremely advantageous property in which precise control over power and optical interactions can be maintained and exploited. Several studies have shown excellent results in treating a variety of cancers [10,11].

[1]SPR is the phenomenon involving the collective oscillations of the materials' free conducting electrons when stimulated by incident light at a particular wavelength (usually chosen to match the resonance mode of the oscillating electrons).

CNTs have also been investigated as a platform for cancer hyperthermia [12] as their electromagnetic optical absorption spectrum is very broad and extends into both the NIR I and NIR II windows, which corresponds to the optical transmission window of biological tissue (i.e., propagation of electromagnetic energy through the tissue is deeper at these wavelengths as absorption from biological molecules, such as hemoglobin, oxyhemoglobin, melanin, and water, is greatly decreased). In addition, further tuning of the photophysical properties of the CNTs can be achieved by changing the CNT chirality, length, and diameter of the CNT.

The use of graphene and graphene oxide in photon-mediated hyperthermia has also received attention, primarily due to its broadband optical and near-infrared absorption properties [13] similar to that of CNTs. One of the first investigations reported on ultrahigh in vivo uptake of PEGylated graphene oxide into breast cancer tumors and was accompanied with photothermal therapy using NIR excitation [14]. Although these results were interesting, the high dosage of graphene used in this study would deem it unfit for clinical application.

Overall, although these techniques offer interesting and novel treatment opportunities for a variety of superficial cancer types, they are still limited by the subsurface penetration depth ($\sim$1–4 mm) of optical and near-infrared light. The use of full-body penetrating radiowaves, with penetration depths up to 20–40 cm, offers a selective advantage over phototherapy techniques, provided that the interactions of RF energy with nanomaterials is enough to instigate sufficient levels of heating.

RF INTERACTIONS WITH AuNPs

The interactions of RF electric field energy with AuNPs as a means to induce targeted hyperthermia have been the subject of a great deal of debate over the last 8 years. Recent work had shown that both CNTs and AuNPs promoted noninvasive thermal hyperthermia in a variety of hepatic and pancreatic cancer cell lines in both in vitro and in vivo conditions (biological work in this field will be further discussed in the proceeding sections). However, when examining the heating rates of AuNPs in aqueous, nonbiological conditions, in order to examine the underlying heating mechanisms, the background heating rates of the ionic buffer solution were originally overlooked and became the basis for critical discussions with several publications demonstrating negligible heating of AuNPs when the background buffer heating was taken into account. A full comprehensive review of the research to-date can be found in the literature [15,16].

It was comprehensively shown however that AuNPs do in fact heat when exposed to an RF electric field (operating at 13.6 MHz at a field strength of $\sim$90 kV/m) [17]. Further, this heating was found to be size-dependent (heating was only observed for AuNPs with diameters ≤ 10 nm); concentration-dependent (minimum Au concentration needed was 500 mg/L); power-dependent; and highly dependent on surface charge and aggregation effects, whereby heating was only observed for citrate-capped AuNPs and was diminished when the AuNPs were aggregated. An overview of these findings is highlighted in Fig. 1.4.

Strict purification and cleaning protocols were utilized to create AuNPs suspended in deionized (DI) water with close to zero background contaminants. An overview of this purification process is shown in Fig. 1.5. Briefly, as-purchased solutions (Ted Pella, Inc., Redding, CA) of citrate-capped AuNPs of diameters 5, 10, 20, and 50 nm were centrifuged (3500 rpm for 125 s) through 50-kDa filters to remove the background buffers. This was repeated several times with the addition of highly purified DI water for continuous removal of background ions and contaminants. The AuNPs were then concentrated and subjected to RF exposure (900 W, 13.56 MHz). Using this technique, we completely eliminated background heating rates from the buffers and were able to examine the heating rates caused primarily by the AuNPs themselves.

As previously stated, the heating mechanism of AuNPs in an RF environment has been the cause for much debate and criticism. The most likely reason that AuNP heating rates were not observed by the work of other groups [18–23] is due to discrepancies in regards to operating power, system setup, AuNP concentrations, and aggregation effects. For example, in the work by Li et al. [22], although they had a similar system to Corr et al. [17], they only subjected their samples to 25 W of RF

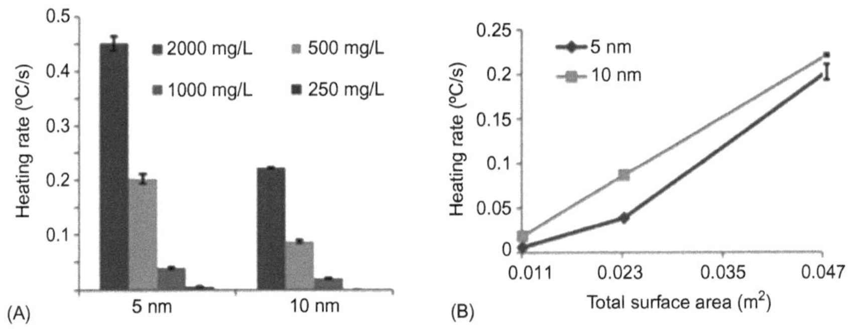

FIGURE 1.4

(A) Heating rates for 5- and 10-nm AuNPs as a function of Au concentration. Background ionic surfactant heating rates have been subtracted from all AuNP heating data. Heating was not observed for AuNPs of diameters 20 and 50 nm. (B) Heating rates versus total AuNP surface area in a 1.3-mL sample (for concentrations 500 – 2000 mg/L). Heating rates for 5- and 10-nm AuNPs are approximately equivalent, indicating a strong surface area relationship to heating.

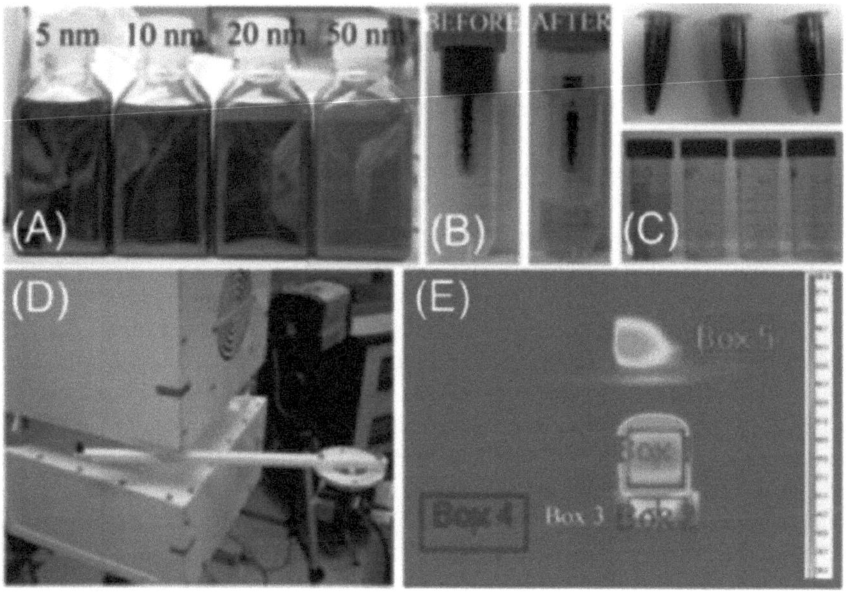

FIGURE 1.5

(A) As-purchased AuNPs of diameter 5–50 nm suspended in their buffers. (B) Solutions were lightly centrifuged (3500 rpm) through a 50-kDa filter for 125 s. The resultant isolated buffer can be seen in the right-hand image. (C) Cleaned AuNPs were then concentrated and resuspended in deionized water (top). Removed buffers (bottom picture) were collected and exposed to RF heating conditions. (D) Samples were positioned on a Teflon sample holder and exposed to the RF field. (E) Temperature changes and heating rates were recorded using an infrared camera.

power, compared to 900 W. They were successful in showing that the background heating rates were responsible for previously observed heating effects [24] but could not elicit heating effects from the AuNPs themselves as they only tested aggregated "pelleted" forms of the AuNPs mainly of diameters 20, 50, and 200 nm, and at a significantly lower operating power and Au concentration.

Similar to this work, Kruse et al. [23] constructed an RF coupling network to investigate the heating of aqueous suspensions of citrate-capped AuNPs purchased from the same supplier as quoted in the work by both Moran et al. [24] and Corr et al. [17]. Although their design was different in that they utilized a "resonant-mode" device rather than the Kanzius "transmission-mode" device, the operating principles were the same in that both systems exposed the AuNPs to a 13.56-MHz electric field of variable power. Similar to previous results, they were able to demonstrate significant heating of the AuNPs suspensions for AuNPs of diameters 5, 10, 20, and 50 nm, but did not take into account the background heating effects of the buffer solutions. Hence, no information about the heating mechanisms of AuNPs in an RF electric field could be obtained.

A good amount of work has also been published by Liu et al. [16,18,20,25]. In one publication [20], AuNPs of size distribution 10–30 nm in diameter were synthesized in-house using the standard citrate reduction method and were subsequently purified somewhat by using a membrane dialysis method over a period of 5 days. The resulting AuNPs were then tested in their RF setup, which consisted of a power generator, a microstrip waveguide, and a matching termination to produce a relatively uniform electric field (~16 kV/m) across a 10-mm air gap, in which the AuNPs were placed. Similar to previous results, they were able to demonstrate slight heating of the buffer suspension but negligible heating of the purified samples. Again, direct observable heating of AuNPs was most likely not seen due to their substantially lower operating power and AuNP concentrations. For example, in the results of Corr et al., background buffer heating rates of 0.5°C/s were observed compared to 0.0015°C/s in the work of Liu et al.

The RF heating effects of aggregated AuNPs were also examined by Liu et al. [18] and compared to their bulk heating behavior under exposure to a 655-nm laser. Although aggregation effects were shown to dramatically increase the heating behaviors for the AuNPs being exposed to optical stimulation, negligible amounts of heat production were evident for the RF exposure tests. In their experiment, AuNPs were synthesized, prepared, and characterized in-house similar to that stated previously. In order to prepare aggregated AuNPs of equal concentration, they exposed various AuNP solutions to a DC electrode configuration under different voltages (0–11 V) for different set times (0–9 min). This technique allowed for various AuNP "clusters" or "aggregates" to form, the size of which varied from 10 to 100 nm. Also, the aggregation state could also be monitored effectively using dynamic light scattering (DLS) as well as monitoring the main plasmonic resonance peak using UV–vis spectroscopy. Although significant changes in temperature of the bulk aqueous solution could be achieved using laser excitation, negligible levels of heat were observed for the RF studies, again, most likely, due to size-, concentration-, power-, and aggregation-effects.

Finally, the effect of sample holder geometry in relation to AuNP heating effects should also be mentioned as is has recently been shown that for linearly polarized electric fields it is important to take into account container shape and orientation when measuring the heating rates of samples [21]. Their work demonstrated that long thin cuvettes (e.g., test tubes) can cause RF absorption to be several orders of magnitude higher than spherical sample holders (e.g., round bottom flasks), that can, in turn, cause much higher absorption than thin disks often used in cell-culture experiments (e.g., Petri dishes). Although the effect of sample holder geometry plays an important effect in the observed heating rates, they were still unable to demonstrate AuNP heating rates primarily due to the AuNP concentration (~50 mg/L) and large size (100 nm diameter) as well as variations in RF power (25 W vs 900 W).

THEORETICAL FRAMEWORKS FOR RF-INDUCED AuNP HEATING

JOULE MODEL

The first theoretical model to be proposed as the main mechanism of RF-induced heating effects in AuNPs was the Joule model: a resistive heating mechanism that models the flow of electrons through conductors (i.e., AuNPs) when exposed to an electric field (E):

$$I^2R = \sigma(\omega)\cdot\left[a\cdot d\cdot|E|^2\right] \tag{1.1}$$

where a is the AuNP cross-sectional area, d is the diameter, and $\sigma(\omega)$ is the frequency-dependent (rad/s) Drude model conductivity. Eq. (1.1) can be incorporated in terms of heating rates by using the following equation:

$$\frac{dT}{dt} = N\cdot\frac{\sigma(\omega)\cdot a\cdot d\cdot|E|^2}{v\cdot C_W} \tag{1.2}$$

where v is the sample volume, N is the number of AuNPs, and C_W is the specific heat capacity of water. This model seemed to predict and model the heating behaviors of the AuNP aqueous solutions shown in the work of Moran et al. [24], but subsequently became invalid as the authors did not take into account the bulk heating rates of the AuNP buffers, as mentioned previously.

CLASSICAL AND QUANTUM ELECTROMAGNETIC THEORY

An excellent theoretical account of heating mechanisms in AuNPs can be found in the work of Hanson et al. [26,27]. In their work, they first investigated the optimal electrical conductivity of nanoparticles that would achieve maximum energy deposition, as well as varying the size and aspect ratio of various nanoparticle configurations (e.g., spheres, cylinders, with and without dielectric coatings). Their model is based on the assumption that in order to heat a nanoparticle with electromagnetic energy, the nanoparticle must have an effective conductivity (σ) whereby all loss mechanisms (i.e., dipolar friction, conduction, etc.) can be described by a finite, nonzero imaginary permittivity (ε''), which is expressed through the equation:

$$\varepsilon = \varepsilon' - j\varepsilon'' = \left(\varepsilon_r' - j\sigma/\omega\varepsilon_0\right)\varepsilon_0 \tag{1.3}$$

where ε' is the real permittivity and ε_0 is the vacuum permittivity of free space. Using this equation as the basis of heating mechanisms, it can clearly be seen that in the case of infinite conductivity (i.e., a perfect conductor) the material wouldn't heat, as there are no loss mechanisms. The same is true for the opposite case (i.e., zero conductivity). Hence there should be an optimal conductivity for a given nanoparticle and shape that allows for maximum RF absorption and heating by

AuNPs. Their results specified that in all cases, a relatively small conductivity value (i.e., $\sigma \ll 1$ S/m) would result in maximum heating, suggesting that lossy dielectrics may be the best nanoparticle material to be used, although this has still to be experimentally determined.

In their second addition [26], Hanson et al. investigated other mechanisms by which nanoparticles can absorb low-frequency electromagnetic radiation, which include both classical and quantum effects. Their classical model integrated the use of permittivity and absorption cross sections into a generalized Mie model of electromagnetic interactions, which also included factors such as electric and magnetic dipole polarizability, as well as use of the Drude formula to calculate frequency-dependent complex permittivity based on the bulk relaxation time and effective plasma frequency (see Ref. [26] for more details). The quantum effects studied include nonlocal electronic surface effects, electron spill-out, and surface roughness. The generalized conclusions were that none of these mechanisms could account for the reported rise in temperature of AuNP suspensions such as those in Ref. [24]. Further, they went so far as to analyze the absorption of RF energy in the host medium, which was (as stated previously) the underlying reason for observed heating effects in the bulk solutions.

Their theoretical observations were in line with the previously mentioned experimental evidence [20,22] in which absorption of RF energy by the host medium is the primary factor for observed RF-induced heating phenomenon. They did however suggest that quantum effects such as those arising from needle-like objects (i.e., CNTs) and elongated ellipsoidal nanoparticles can maintain a strong and localized, radial near-field in the vicinity of the nanoparticle, which in turn can significantly enhance absorption in a conductive host. These statements remain, as of yet, unproved although recent work evaluating the heating behaviors of highly enriched semiconducting and metallic single-walled CNTs in an RF field demonstrate both supporting and contradicting results [28].

ELECTROPHORETIC MODEL

It was shown theoretically [29], and strongly supported indirectly by experimental evidence [17], that the electrophoretic movement of charged particles in response to a time-varying electric field may be a mechanism of great importance. This makes a great deal of sense in regards to AuNP solutions given that the citrate-charge capping layer on the AuNP surface is negatively charged (zeta potential ~ -20 to -40 mV) and would respond to a time-varying electric field described by:

$$E = E_0 e^{j\omega t} \tag{1.4}$$

where t is time (s). The movement of these particles in relation to this time-varying electric field can be expressed by the following equation:

$$m\frac{dv}{dt} + \beta v = qE \tag{1.5}$$

where m is the nanoparticle mass, v is the velocity of the nanoparticle through the medium, q is the particle charge, β is the friction constant $\beta = 6\pi\mu_f a$ (where a is the nanoparticle diameter and μ_f is the viscosity coefficient of the host medium). Further integration of other terms (see Ref. [29] for full derivation) led to the following important equations:

$$\varepsilon' = -\frac{Nq^2}{\beta\varepsilon_0}\frac{\tau}{1+(\omega\tau)^2} \tag{1.6}$$

$$\varepsilon'' = \frac{Nq^2}{\beta\omega\varepsilon_0}\frac{\tau}{1+(\omega\tau)^2} \tag{1.7}$$

where $\tau = m/\beta$ and ω is the operating frequency (rad/s). Hence, as can be seen by Eqs. (1.6) and (1.7), the effect of electrophoretic motion placed upon AuNPs when exposed to a time-varying electric field can directly be mathematically incorporated into the effective dielectric permittivity, which can easily be measured by using an impedance analyzer such as the Agilent E4991A. Ultimately, this model suggests that AuNPs act like nanoscale oscillators, moving back and forth in response to a time-varying electric field, generating heat via frictional mechanisms. Corr et al. [17] estimated this oscillating distance to be ~0.1–0.2 Å for each period of a 13.56-MHz operating frequency.

BIOLOGICAL RF ACTIVITY OF AuNPs IN VITRO/IN VIVO

Although there has been a great deal of work centered around understanding the heating mechanism of AuNPs to optimize and approve on their design so that they can better target and induce hyperthermia in living, biological systems, there has actually been limited translation into in vitro and in vivo investigations primarily due to the lack of institutes and laboratories with the necessary RF equipment. Most of the work has come out of the Curley laboratories with specific examples centered around the stability of antibody-conjugated AuNPs in the endolysosomal nanoenvironment [30] as well as AuNPs and RF as experimental models for hepatocellular carcinoma [31].

The work of Raoof et al. [30] demonstrated that liver cells selectively internalize antibody conjugated AuNPs (C225, Cetuximab) through receptor-mediated endocytosis, with the nanoparticles predominantly accumulating and forming aggregates within cytoplasmic endolysosomes. Further, they demonstrated that RF-induced cytotoxicity (induced by RF dissipation and subsequent heating within the cell) was dependent on the colloidal stability of the AuNPs within the endolysosomes, which is pH dependent. This supports previous claims that aggregated AuNPs in aqueous solutions do not heat [17]. They could circumvent this limitation however by administering lysosomotropic agents (e.g., chloroquine, concanamycin A, and ammonium chloride) to modulate the endolysome pH, preventing AuNP aggregation and increasing RF-induced cytotoxicity. Highlights from this work can be seen in Fig. 1.6.

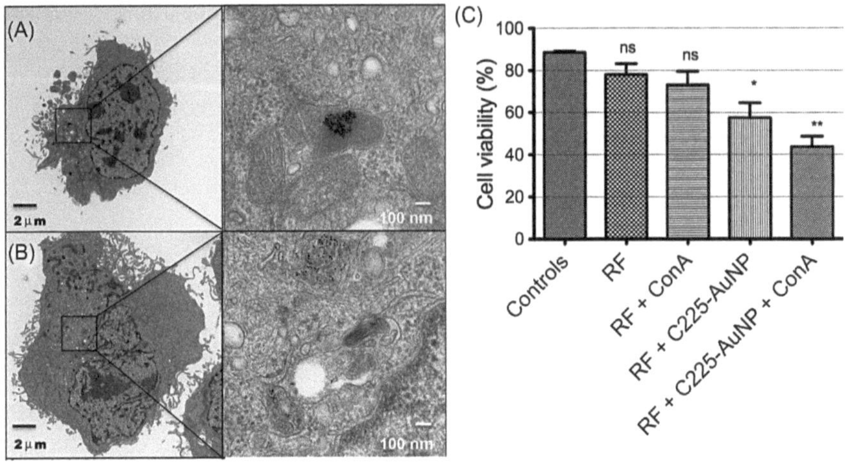

FIGURE 1.6

Intracellular localization and aggregation state of AuNPs imaged by transmission electron microscopy (TEM). SNU449 cells were incubated with C225-AuNPs (100 μg/mL) in the absence (A) or presence (B) of concanamycin A (10 nM). (C) Hyperthermic cytotoxicity after RF field exposure (13.56 MHz, 600 W) treatment. SNU449 cells were incubated with C225-AuNP (200 μg/mL) in the absence or presence of concanamycin A (10 nM) for 4 h immediately followed by RF exposure (9 min). Cell viability was assessed by flow cytometry–based assay (Annexin-V & PI staining) 24 h after RF exposure. **p*-value <0.05, ***p*-value <0.01.

The use of targeted, drug-AuNP conjugates in combination with noninvasive RF hyperthermia was also recent investigated by Raoof et al. [31]. In this work, AuNPs (10 nm diameter) were conjugated to C225 (Cetuximab[2]) and were loaded with GCB using methoxypoly(ethylene glycol)-thiol (mPEG-SH) to create an AuNP-C225-GCB-mPEG conjugate. These targeted nanoconjugates, in combination with RF therapy, were selectively cytotoxic to epidermal growth factor receptor-expressing liver cancer cell lines (Hep3B and SNU449) in vitro when compared to isotype nanoconjugates (IgG) with and without RF. Further to this, in vivo experiments with subcutaneous Hep3B xenografts halted the growth of these tumors as well as demonstrating increased apoptosis, necrosis, and decreased proliferation rates. Highlights from these results are shown in Fig. 1.7. When evaluating the concentration of GCB used in the AuNP nanoconjugates, when compared to the nonconjugated GCB, it was determined that targeted delivery of GCB is more effective and safe at dosages $\sim 275\times$ lower than the current clinically administered systemic dose of GCB. One possible mechanism for this observed synergy is from the ability of RF hyperthermia to inhibit recombination repair mechanisms of GCB-stalled replication forks [32].

[2]Cetuximab is an epidermal growth factor inhibitor antibody, commonly used for the treatment of metastatic colorectal cancer, metastatic nonsmall cell lung cancer, and a variety of head and neck cancer.

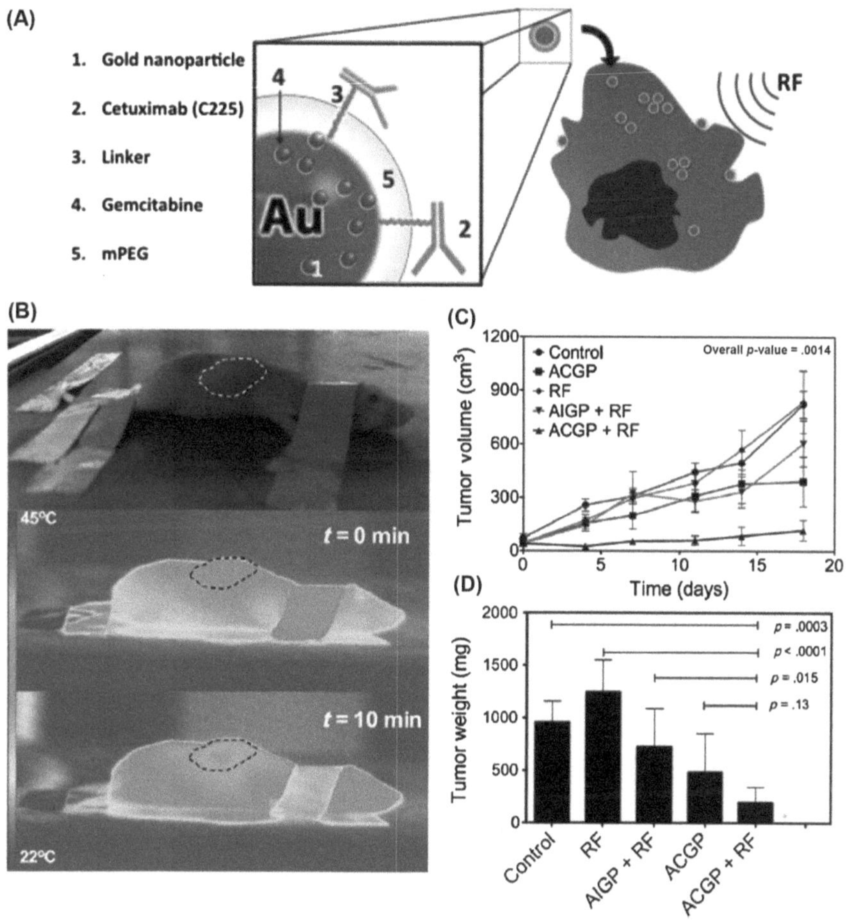

FIGURE 1.7

(A) Schematic of AuNP-C225-GCB-mPEG conjugate used for targeting RF-induced hyperthermia in hepatocellular cancer cells. (B) A BalbC/Nu mouse is anesthetized, injected with conjugated AuNPs, and placed in the RF generator (shown previously in Fig. 1.2). (C) Changes in tumor volume are recorded after the start of treatment at Day 0. ACGP are the C225-targeted AuNP conjugates while AIGP are the control (IgG) antibody-conjugated AuNPs. (D) Tumor mass was documented 24 h after last RF treatment or 48 h after last nanoparticle treatment. As can be seen, there is a significant decrease in tumor size and mass for the mice treated with the AuNP-C225-GCB-mPEG conjugated with RF when compared to the control groups.

CONCLUSION

The use of AuNPs as targeted, heating agents are currently being explored in the field of noninvasive RF cancer hyperthermia. It is hypothesized that targeted AuNPs, once internalized within cancer cells, will release heat when activated by exposure to an external noninvasive RF electric field, causing necrosis and apoptosis only in cells that have internalized the AuNPs.

Much work has been focused on understanding the heating mechanism by which AuNPs heat in an RF field in order to fully optimize and maximize RF-induced heating. Although there was initially intense speculation and criticism on whether AuNPs were "hot or not" when exposed to an RF field, recent work has indeed shown that AuNPs do, in fact, heat in an RF field in a concentration-, size-, surface charge-, and power-dependent manner. This can only be observed however if specific protocols are maintained to remove the background buffers and residual contaminants, which have shown to overshadow AuNP heating effects. Also, from several theoretical investigations, which looked at classical and quantum effects such as traditional Mie absorption/scattering and electron confinement effects, it seems that electrophoretic heating is the most likely mechanism for observed effects.

When translated into in vitro and in vivo experiments, AuNPs have shown interesting results and can cause an increase in RF-induced cellular cytotoxicity provided that the AuNPs remain nonaggregated. As they primarily accumulate in the cellular endolysosomes, the use of lysosomotropic agents to modulate the pH can prevent aggregation and enhance toxicity. Complete tumor suppression has also been shown in hepatocellular models of cancer using an AuNP-drug-antibody conjugate that specifically targets EGFR. Further, the use of GCB in the conjugation was shown to greatly enhance the effects of the drug as dosages $\sim 275\times$ lower than what is systemically delivered in the clinic today. Given these findings, the use of AuNPs in noninvasive RF cancer hyperthermia is still an area of active research.

REFERENCES

[1] Roussakow S. The history of hyperthermia rise and decline. Conf Papers Med 2013;2013:40.

[2] Abe M, Hiraoka M, Takahashi M, Egawa S, Matsuda C, Onoyama Y, et al. Multi-institutional studies on hyperthermia using an 8-MHz radiofrequency capacitive heating device (thermotron RF-8) in combination with radiation for cancer therapy. Cancer 1986;58:1589–95.

[3] Issels RD, Lindner LH, Verweij J, Wust P, Reichardt P, Schem B-C, et al. Neo-adjuvant chemotherapy alone or with regional hyperthermia for localised high-risk soft-tissue sarcoma: a randomised phase 3 multicentre study. Lancet Oncol 2010;11:561–70.

[4] Corr SJ, Shamsudeen S, Vergara LA, Ho JC-S, Ware MJ, Keshishian V, et al. A new imaging platform for visualizing biological effects of non-invasive radiofrequency electric-field cancer hyperthermia. PLoS ONE 2015;10:e0136382.
[5] Huang X, El-Sayed MA. Gold nanoparticles: optical properties and implementations in cancer diagnosis and photothermal therapy. J Adv Res 2010;1:13–28.
[6] Dresselhaus MS, Dresselhaus G, Saito R. Physics of carbon nanotubes. Carbon N Y 1995;33:883–91.
[7] Sethi R, Mackeyev Y, Wilson LJ. The Gadonanotubes revisited: a new frontier in MRI contrast agent design. Inorgan Chim Acta 2012;393:165–72.
[8] Sethi R, Ananta JS, Karmonik C, Zhong M, Fung SH, Liu X, et al. Enhanced MRI relaxivity of Gd3 + -based contrast agents geometrically confined within porous nanoconstructs. Contrast Media Mol Imaging 2012;7:501–8.
[9] Kennedy LC, Bickford LR, Lewinski NA, Coughlin AJ, Hu Y, Day ES, et al. A new era for cancer treatment: gold-nanoparticle-mediated thermal therapies. Small 2011; 7:169–83.
[10] Strong LE, West JL. Optically modulated cancer therapeutic delivery: past, present and future. Ther Deliv 2015;6:545–58.
[11] Ashton JR, Clark DP, Moding EJ, Ghaghada K, Kirsch DG, West JL, et al. Dual-energy micro-CT functional imaging of primary lung cancer in mice using gold and iodine nanoparticle contrast agents: a validation study. PLoS ONE 2014;9: e88129.
[12] Singh R, Torti SV. Carbon nanotubes in hyperthermia therapy. Adv Drug Deliv Rev 2013;65:2045–60.
[13] Orecchioni M, Cabizza R, Bianco A, Delogu LG. Graphene as cancer theranostic tool: progress and future challenges. Theranostics 2015;5:710–23.
[14] Yang K, Zhang S, Zhang G, Sun X, Lee S-T, Liu Z. Graphene in mice: ultrahigh in vivo tumor uptake and efficient photothermal therapy. Nano Lett 2010;10: 3318–23.
[15] Collins CB, McCoy RS, Ackerson BJ, Collins GJ, Ackerson CJ. Radiofrequency heating pathways for gold nanoparticles. Nanoscale 2014;6:8459–72.
[16] Liu X, Chen H-j, Chen X, Alfadhl Y, Yu J, Wen D. Radiofrequency heating of nanomaterials for cancer treatment: progress, controversies, and future development. Appl Phys Rev 2015;2:011103.
[17] Corr SJ, Raoof M, Mackeyev Y, Phounsavath S, Cheney MA, Cisneros BT, et al. Citrate-capped gold nanoparticle electrophoretic heat production in response to a time-varying radio-frequency electric field. J Phys Chem C 2012;116:24380–9.
[18] Liu X, Chen H-j, Chen X, Alfadhl Y, Yu J, Wen D. Electromagnetic heating effect of aggregated gold nanoparticle colloids. J Appl Phys 2014;115:094903.
[19] Kim HK, Hanson GW, Geller DA. Are gold clusters in RF fields hot or not? Science 2013;340:441–2.
[20] Liu X, Chen H-j, Chen X, Parini C, Wen D. Low frequency heating of gold nanoparticle dispersions for non-invasive thermal therapies. Nanoscale 2012;4:3945–53.
[21] Dongxiao L, Yun Suk J, Hong Koo K, Junda C, Geller DA, Shuba MV, et al. The effect of sample holder geometry on electromagnetic heating of nanoparticle and NaCl solutions at 13.56 MHz. Biomed Eng, IEEE Trans 2012;59:3468–74.
[22] Li D, Jung YS, Tan S, Kim HK, Chory E, Geller DA. Negligible absorption of radiofrequency radiation by colloidal gold nanoparticles. J Colloid Interface Sci 2011;358: 47–53.

[23] Kruse DE, Stephens DN, Lindfors HA, Ingham ES, Paoli EE, Ferrara KW. A radio-frequency coupling network for heating of citrate-coated gold nanoparticles for cancer therapy: design and analysis. Biomed Eng, IEEE Trans 2011;58:2002–12.

[24] Moran C, Wainerdi S, Cherukuri T, Kittrell C, Wiley B, Nicholas N, et al. Size-dependent joule heating of gold nanoparticles using capacitively coupled radio-frequency fields. Nano Res 2009;2:400–5.

[25] Liu X, Chen H-J, Alfadhl Y, Chen X, Parini C, Wen D. Conductivity and frequency dependent specific absorption rate. J Appl Phys 2013;113:074902.

[26] Hanson GW, Monreal RC, Apell SP. Electromagnetic absorption mechanisms in metal nanospheres: bulk and surface effects in radiofrequency-terahertz heating of nanoparticles. J Appl Phys 2011;109:124306.

[27] Hanson GW, Patch SK. Optimum electromagnetic heating of nanoparticle thermal contrast agents at RF frequencies. J Appl Phys 2009;106:054309–10.

[28] Corr S, Raoof M, Cisneros B, Orbaek A, Cheney M, Law J, et al. Radiofrequency electric-field heating behaviors of highly enriched semiconducting and metallic single-walled carbon nanotubes. Nano Res 2015;8:2859–70.

[29] Sassaroli E, Li KCP, Neill BEO. Radio frequency absorption in gold nanoparticle suspensions: a phenomenological study. J Phys D Appl Phys 2012;45:075303.

[30] Raoof M, Corr SJ, Kaluarachchi WD, Massey KL, Briggs K, Zhu C, et al. Stability of antibody-conjugated gold nanoparticles in the endolysosomal nanoenvironment: implications for noninvasive radiofrequency-based cancer therapy. Nanomed Nanotechnol Biol Med 2012;8:1096–105.

[31] Raoof M, Corr SJ, Zhu C, Cisneros BT, Kaluarachchi WD, Phounsavath S, et al. Gold nanoparticles and radiofrequency in experimental models for hepatocellular carcinoma. Nanomed Nanotechnol Biol Med 2014;10:1121–30.

[32] Raoof M, Zhu C, Cisneros BT, Liu H, Corr SJ, Wilson LJ, et al. Hyperthermia inhibits recombination repair of gemcitabine-stalled replication forks. J Natl Cancer Inst 2014;106.

CHAPTER

Silk fibroin nanoparticles and cancer therapy

2

D. Mishra, T.S. Iyyanki, J.R. Hubenak, Q. Zhang and A.B. Mathur
The University of Texas MD Anderson Cancer Center, Houston, TX, United States

CHAPTER OUTLINE

INTRODUCTION

Despite many recent advances in breast cancer treatment, breast cancer is the second leading cause of cancer death in women in the United States [1]. Although only 5% of breast cancer diagnoses are classified as metastatic disease, which has a 23.8% survival rate within 5 years of detection, nearly 30% of women diagnosed with earlier stages develop into metastatic disease [2,3]. Common treatment options for breast cancer include chemotherapy, hormone/endocrine therapy, and radiation [4,5]. However, these therapies have very low survival rates and are less effective with metastatic disease. Thus there is a need to improve the treatment of metastatic diseases.

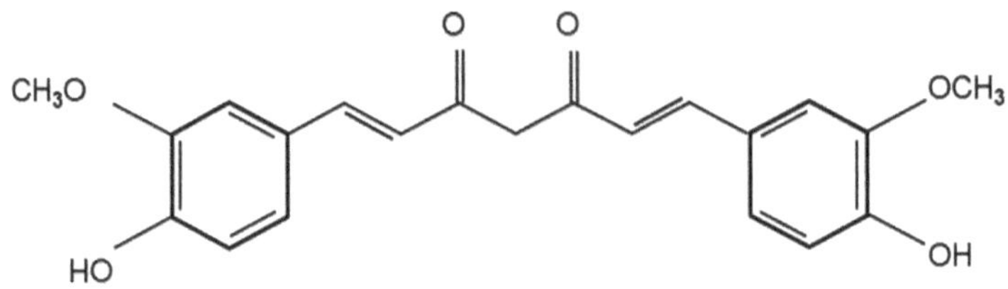

FIGURE 2.1

The chemical structure of curcumin.

Curcumin is a natural drug derived from the East Indian plant *Curcuma longa* (Fig. 2.1). It has been shown that curcumin possesses antioxidant, antiinflammatory, and antiproliferative properties, giving it the potential to treat a myriad of diseases including Alzheimer's, Parkinson's, and cancer [6–8]. As a cancer treatment, curcumin can interact and bind directly to various proteins, causing the suppression of several signaling pathways which then inhibits angiogenesis, cancer cell proliferation, and metastasis. While the potential usage of curcumin to treat various diseases including cancer is currently being evaluated in clinical trials [9], preclinical studies in animal models have demonstrated curcumin has chemopreventive properties in treating many different cancers including breast, lung, kidney, bladder, blood, skin, colon, intestinal, esophageal, stomach, oral, and liver cancers [10,11]. Due to its hydrophobicity, curcumin is typically administered through either oral or intraperitoneal routes. However, curcumin bioavailability is very low due to poor absorption, rapid metabolism, low serum levels, and low tissue distribution [12]. Hence there is a need to increase systemic distribution and bioavailability of curcumin for better preclinical and clinical outcomes.

Drug transport to a specific anatomical location and local tissue penetration upon delivery are major problems facing the translation of chemotherapeutic drugs into the clinic [12]. To address this, drug delivery systems such as micelles, liposomes, and nanoparticles are being developed and evaluated for transporting chemotherapeutic agents directly to tumor tissues [13–20]. In addition to aiding therapeutic transport, nanoparticle drug delivery systems allow controlled drug release and improve drug absorption and distribution. Clinically approved nanoparticle-based drug delivery systems are currently used to treat several diseases including hepatitis, multiple sclerosis, and breast cancer [21]. Albumin-bound paclitaxel nanoparticles is a clinically approved delivery system that was shown to increase drug bioavailability, drug accumulation in tumor, and improve patient outcomes [22]. Due to these advantages, there has been increased interest in developing polymer-based drug delivery systems.

Silk fibroin (SF) is a naturally derived protein from the *Bombyx mori* silk worm (Fig. 2.2) that has been used by our group to develop nanoparticles for drug delivery and manufacture scaffolds for tissue engineering and wound healing applications. Our previous work established the potential of using SF in the

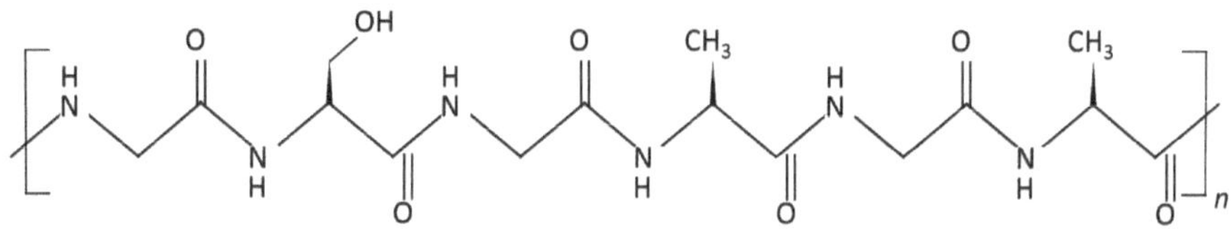

FIGURE 2.2

The chemical structure of silk fibroin.

encapsulation and controlled release of emodin from liposomes and curcumin drug for the treatment of breast cancer cell lines. However the effect of curcumin and the curcumin-loaded SF nanoparticles has not been well studied in a highly progressed metastatic breast cancer cell line.

In this study, we prepared curcumin-loaded SF nanoparticles using solvent precipitation techniques and evaluated their efficacy in inducing apoptosis in a metastatic breast cancer cell line in vitro.

MATERIALS AND METHODS

MATERIALS

DMEM/F12 cell culture medium, bovine serum albumin, monoclonal antiβ-actin antibody, sodium chloride, sodium deoxycholate, sodium bicarbonate, and 3-(4,5-dimethyl-2-thiazolyl)-2,5-diphenyl-2*H*-tetrazolium bromide (MTT) were purchased from Sigma-Aldrich Co. (St Louis, MO, USA). Fetal bovine serum (FBS), penicillin/streptomycin, and 0.25% trypsin/EDTA were purchased from Gibco BRL (Grand Island, NY, USA). Dithiothreitol (DTT), calcium nitrate tetrahydrate, glycine, methanol, Triton-X 100, Tween-20, dimethyl sulfoxide (DMSO), SnakeSkin* dialysis tubing MWCO 3.5 kDa were purchased from Thermo Fisher Scientific (Waltham, MA, USA). Coomassie (Bradford) protein assay kit was purchased from Bio-Rad Laboratories (Hercules, CA, USA). Ca^{2+}-free and Mg^{2+}-free phosphate-buffered saline (PBS) was purchased from Mediatech, Inc. (Manassass, VA, USA).

CELL CULTURE

GILM2 cells (human lung metastatic-derived breast cancer cell line, kind gift from Dr Janet E. Price, MD Anderson Cancer Center, Houston, TX, USA) were used to determine the cytotoxicity of free curcumin, SF nanoparticles, and curcumin-loaded SF nanoparticles, and to isolate protein for the protein array experiments. For all experiments, GILM2 cells were grown in DMEM/F12 medium supplemented with 10% heat-inactivated FBS, 1% penicillin/streptomycin, and 1% insulin/transferrin/selenium (Corning Cellgro, Manassas, VA, USA) at 37°C in humidified air containing 5% CO_2.

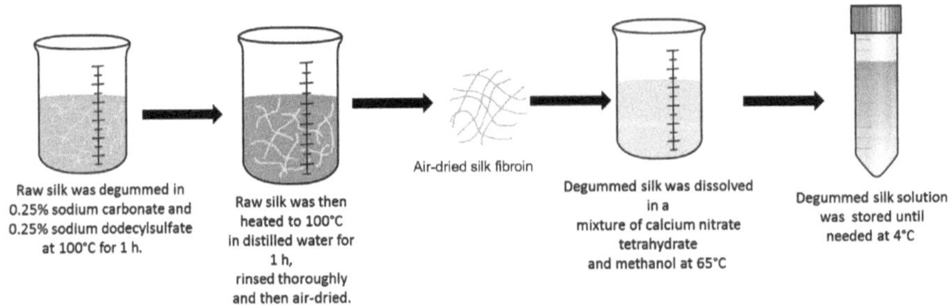

FIGURE 2.3

A schematic showing the preparation of silk fibroin solution.

PREPARATION OF SF SOLUTION AND CURCUMIN SOLUTION

SF solution was prepared as described in Fig. 2.3 [23]. Briefly, raw silk (kind gift from Dr Hudson, North Carolina State University, Raleigh, NC, USA) was degummed in 0.25% (w/v) sodium carbonate and 0.25% (w/v) sodium dodecylsulfate at 100°C for 1 hour. Raw silk was then heated in distilled water at 100°C, rinsed thoroughly in running distilled water, and air-dried. The degummed silk was dissolved in a solution of calcium nitrate tetrahydrate and methanol (molar ratio of 1:4:2 Ca:H_2O:MeOH at 65°C) and stirred for 3 hours. SF solution was dialyzed (MWCO 3.5 kDa) against distilled water for 48 hours. The dialyzed solution was filtered with 0.22-μm filter and then diluted with double distilled deionized water (ddH_2O) to 0.05% (w/v) for nanoparticle preparation. Concentration of the silk solution was verified by Bradford protein assay. Curcumin solution was prepared by dissolving curcumin (ChromaDex, Inc., Irvine, CA, USA) in 200 proof ethanol at 0.1 mg/mL, which was stored at −80°C until use.

PREPARATION OF SF NANOPARTICLES AND CURCUMIN-LOADED SF NANOPARTICLES

SF and curcumin-loaded SF nanoparticles were prepared using solvent precipitation (Fig. 2.4). SF solution (prepared as described in Fig. 2.3) was mixed with ethanol at a ratio of 6:4 (v/v). For example, 6 mL of 0.05% (w/v) SF solution was mixed with 4 mL of ethanol (EtOH) or 4 mL of 0.1 mg/mL curcumin-ethanol in 15-mL falcon tubes. The mixture was briefly vortexed, flash-frozen in liquid nitrogen, and allowed to thaw at room temperature. Once thawed, samples were centrifuged at 14,000 rpm for 5 minutes. The supernatant was discarded and the pellet was collected and washed with ethanol to crystallize the SF and remove any unbound curcumin. After further centrifugation at 14,000 rpm for 5 minutes, the supernatant was discarded. The remaining pellet was resuspended in double

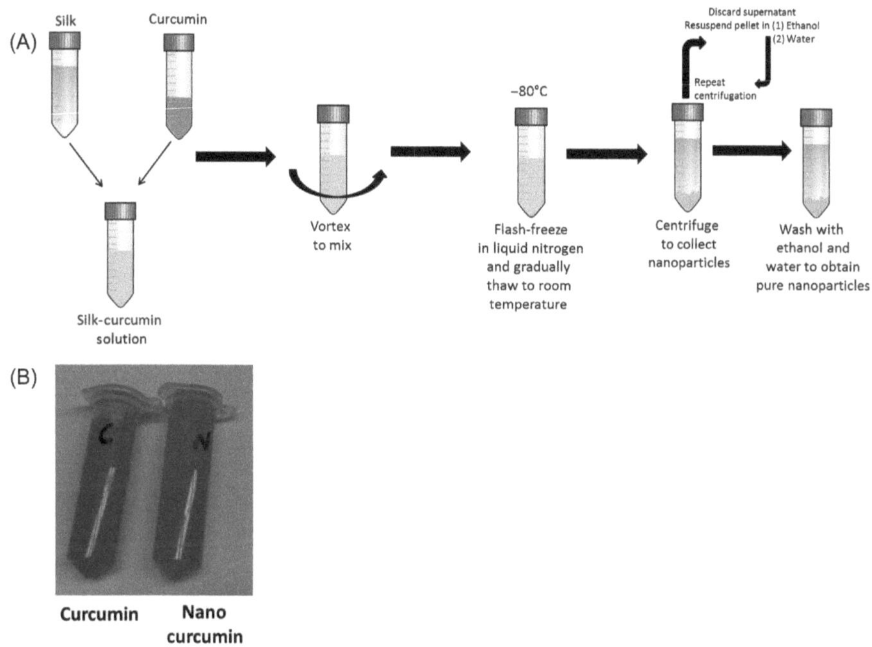

FIGURE 2.4

(A) A schematic showing the nanoparticle fabrication. Nanoparticles are formed by adding curcumin-ethanol solution (0.1 mg/mL at −80°C) to 0.05% SF (2:3 EtOH:SF) ratio. The solution is vortexed then flash-frozen in liquid nitrogen. Once thawed, the solutions are centrifuged at 3700 × *g*, rinsed with pure ethanol, and resuspended in ddH_2O. SF-curcumin nanoparticles suspended in ddH_2O remain suspended for weeks/months and once centrifuged easily resuspend with minor vortexing. (B) Side-by-side comparison of pure curcumin (C) dissolved in DMSO/H_2O and SF-Curcumin (N) nanoparticles. Both samples contain ~10 mg curcumin in 2 mL of solution. The nanoparticle solution appears lighter in color due to the curcumin being entrapped by a silk fibroin particle coating. These are pictures of samples used for in vivo injection.

distilled water and centrifuged at 14,000 rpm twice to wash away any residual ethanol. The nanoparticles were resuspended in 1 mL of 1 × PBS and used for further studies.

CHARACTERIZATION OF CURCUMIN-LOADED SF NANOPARTICLES

Scanning electron microscopy

SF nanoparticles and curcumin-loaded SF nanoparticles suspended in double distilled water were mounted on double-stick carbon tabs (Ted Pella Inc., Redding, CA, USA) and attached to aluminum specimen mounts (Electron Microscopy

Sciences, Ft. Washington, PA). The samples were coated with platinum alloy to a thickness of 25 nm under vacuum using a Balzer MED 010 evaporator (Technotrade International, Manchester, NH), then immediately flash carbon coated under vacuum. Samples were examined using a JSM-5910 scanning electron microscope (JEOL, USA, Inc., Peabody, MA, USA) at an accelerating voltage of 5 kV.

Atomic force microscopy

SF nanoparticles and curcumin-loaded SF nanoparticles were prepared as described above and suspended in water as the final step. Nanoparticle solutions were spread on a glass slide and allowed to air dry at room temperature for 24 hours. ScanAsyst Air AFM tips ($k = 0.4$ N/m; Bruker Corporation, Santa Barbara, CA, USA) were used to scan nanoparticles and obtain height-map images. Particle diameters were measured using ImageJ software (NIH).

Dynamic light scattering

SF nanoparticles with and without curcumin were prepared in ddH_2O and diluted 1:100 for particle size analysis using dynamic light scattering (HORIBA LB-550, Horiba Ltd, Irvine, CA, USA). Particle size distribution measurements were taken in triplicate and averaged together. SF solution alone was analyzed as a control to determine the hydrodynamic size of SF liquid crystals at several concentrations ranging from 0.01% (w/v) to 10% (w/v).

Drug loading efficiency

SF nanoparticles with or without curcumin were prepared as described above with slight modifications. Silk solution (0.6 mL) was mixed with 0.4 mL of ethanol (EtOH) (either with or without 0.1 mg/mL curcumin) in 2 mL Eppendorf centrifuge tubes. The mixture was flash-frozen in liquid nitrogen and thawed at room temperature. Samples were centrifuged at 14,000 rpm for 5 minutes. The supernatant was collected and stored at −20°C for further studies. Supernatant samples were analyzed at 590 nm using UV spectroscopy (MRX Plate Reader, Dynex Technologies, Chantilly, VA, USA) to quantify free curcumin content. Percent drug entrapment is reported as the weight of encapsulated curcumin per milligram of SF nanoparticles.

Curcumin release profile from SF nanoparticles

One milligram of curcumin-loaded SF nanoparticles was suspended in 1 mL of PBS in 2-mL Eppendorf tubes and incubated at 37°C in air containing 5% CO_2 and 90% humidity. At predetermined time points (1, 4, 7, and 10 days incubation), ethanol (1 mL) was added to each tube to dissolve any free curcumin released from the nanoparticles. Nanoparticle solutions were centrifuged at 14,000 rpm for 5 minutes and the curcumin content within the supernatant was measured at 590 nm using UV spectroscopy. Curcumin standards were prepared in a solution of 50% ethanol in water and used for reference.

BIOLOGICAL EVALUATION OF CURCUMIN-LOADED SF NANOPARTICLES

Curcumin efficacy against GILM2 metastatic breast cancer cells

Cytotoxicity of the materials was assessed using standard MTT assay protocols. GILM2 cells were seeded in 96-well tissue culture plates at a density of 5000 cells/well in 0.1 mL of culture medium and incubated for 24 hours. Sample solutions (free curcumin, silk nanoparticles, or curcumin-loaded silk nanoparticles) at various concentrations (0.1 mL) were added to each well and cells were incubated for 24 or 96 hours at 37°C in a cell culture incubator. Curcumin concentrations were calculated based on the amount of curcumin loaded in the nanoparticles such that the delivered dose was equivalent to the free curcumin dose. MTT solution (20 μL, 5 mg/mL) was added to each well and incubated for 4 hours at 37°C in a cell culture incubator. Medium was removed completely, 100 μL of DMSO was added to each well to dissolve the formazan metabolites and the plates were incubated at 37°C for 10 minutes. Sample absorbance was measured at 570 nm using a microplate reader (MRX Reader, Dynex Technologies, Chantilly, VA, USA) and used to calculate cell viability.

Protein expression analysis using protein array

GILM2 cells were seeded in six-well tissue culture plates at a density of 1×10^6 cells/well in 2 mL of culture medium and incubated for 24 hours. Thirty minutes prior to the experiment, the culture medium was replaced with fresh culture medium at a fixed volume of 2 mL per well. Cells were treated with various concentrations of curcumin, silk nanoparticles, or curcumin-loaded silk nanoparticles ($n = 3$ per condition) and incubated for 24 hours at 37°C in a cell culture incubator. To collect the protein, cells were carefully rinsed with 2 mL of cold 1× PBS and harvested by scraping in 300 μL of RIPA buffer. Cells were homogenized for 30 seconds using a probe homogenizer and then centrifuged at 14,000 rpm for 5 minutes at 4°C. Supernatant was collected and pooled ($n = 3$) to create one sample per condition, and protein concentration was quantified using Bradford and BCA assays. Samples were prepared at a concentration of 1 μg/μL (total volume 40 μL) and delivered to the reverse phase protein array core facility at MD Anderson for protein array analysis. At the core facility, the samples were serial diluted five times and arrayed on a nitrocellulose coated slide, probed with 171 antibodies (Table 2.1) using CSA amplication and visualized using DAB colorimetric reaction. Slides were scanned and images were analyzed to quantify density using MicroVigene. Relative protein levels were determined by interpolation from each dilution curve obtained from the standard curve of the slide. Data was normalized for protein loading and transformed into $\log_2$ values, which were used for identifying expression patterns. Relationships between protein expression and potential molecular pathways were analyzed using Ingenuity Pathway Analysis (IPA) software (www.ingenuity.com; Ingenuity Systems, Redwood City, CA, USA). Prior to analysis, the data were normalized against DMSO for free curcumin, and unloaded

Table 2.1 Complete List of Genes Representing the Proteins Investigated Using the Protein Array

ACACA	CTNNA1	MAP2K1	RAD50
ACACB	CTNNB1	MAPK1	RAD51
ACVRL1	DIABLO	MAPK8	RAF1
AKT1	DVL3	MAPK9	RB1
AKT1S1	EEF2	MAPK14	RBM15
AKT2	EEF2K	MET	RICTOR
AKT3	EGFR	MGMT	RPS6
ANXA1	EIF4E	MSH2	RPS6KA1
ANXA7	EIF4EBP1	MSH6	RPS6KB1
AR	EIF4G1	MYC	RPTOR
BAD	ERBB2	MYH11	SCD
BAX	ERBB3	NDRG	SETD2
BCL2	ERCC1	NF2	SFRS1
BCL2L1	ERRFI1	NFkB1	SMAD1
BCL2L11	ESR1	NOTCH1	SMAD3
BECN1	FASN	NRAS	SMAD4
BID	FN1	PARK7	SNAI2
BIRC2	FOXM1	PARP1	SRC
BRAF	FOXO3	PCNA	STAT3
C12orf5	FRAP1	PDCD4	STAT5A
CASP7	G6PD	PDK1	STMN1
CASP8	GAB2	PEA15	SYK
CAV1	GAPDH	PECAM1	TFRC
CCNB1	GATA3	PGR	TGM2
CCND1	GSK3A	PIK3CA	TP53
CCNE1	GSK3B	PIK3R1	TP53BP1
CDC2	IGFBP2	PRKAA1	VHL
CDH1	INPP4B	PRKCA	WWTR1
CDH2	IRS1	PRKCB	XRCC1
CDKN1B	ITGA2	PRKCD	YAP1
CHEK1	KDR	PTEN	YBX1
CHEK2	KIT	PXN	YWHAB
CLDN7	KRAS	RAB11A	YWHAE
COL6A1	LCK	RAB25	YWHAZ

silk nanoparticles for curcumin-loaded nanoparticles to remove the effects of the vehicle. Data were filtered using a log ratio cut-off of 0.15 and then analyzed for expression patterns within various canonical pathways. Based on our data, molecules that correlated strongly with a specific canonical pathway were grouped together, and then further evaluated for the effect of free curcumin and curcumin nanoparticle treatment within that pathway.

Statistics

The statistical significance of the data was evaluated by conducting unpaired Student's t-test with a confidence level of $p < .05$, one-variable Analysis of Variance (ANOVA).

RESULTS

CHARACTERIZATION OF CURCUMIN-LOADED SF NANOPARTICLES

Morphology of curcumin-loaded SF nanoparticles

Scanning electron microscopy was used to identify the morphological structure of curcumin-loaded SF nanoparticles. As seen in Fig. 2.5A, curcumin-loaded SF nanoparticles were uniformly spherical, and aggregated into larger grape-like clusters. Dynamic light scattering measurements confirmed the presence of aggregated particles, giving an average particle diameter of 682.9 ± 177.5 nm (Fig. 2.5B).

Particle size distribution of curcumin-loaded SF nanoparticles

Curcumin-loaded SF nanoparticles were probed using atomic force microscopy to obtain particle size diameter measurements. Although the nanoparticles appeared to be flattened discoid shapes in the image (Fig. 2.6A), this was due to sample preparation methods as particles were observed to be spherical in SEM images. The average particle diameter was determined to be 127.7 ± 6.8 nm (mean ± standard error of mean) as calculated from height-map sections (Fig. 2.6B). As shown in Fig. 2.6C, nanoparticles had a bimodal distribution, with a majority of the particles falling within the 70–110 nm diameter range.

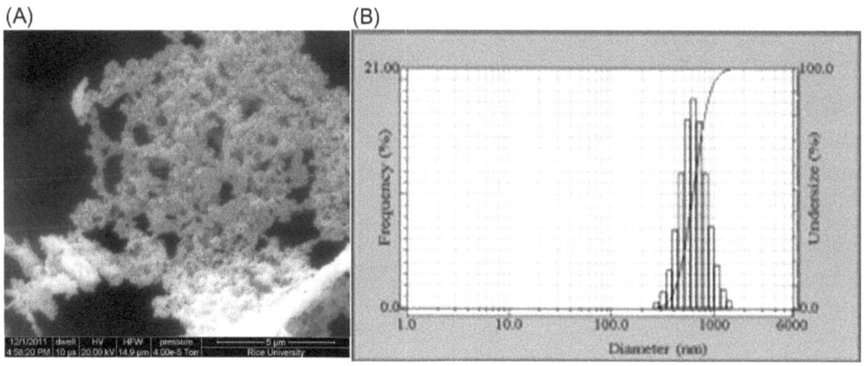

FIGURE 2.5

(A) Scanning electron micrograph of SF-curcumin nanoparticles. Individual nanoparticle morphology is spherical, and nanoparticles appear to have a tendency to aggregate into grape-like clusters. (B) Bar chart showing particle size distribution obtained using dynamic light scattering (DLS). The average particle size was found to be 682.9 ± 177.5 nm, confirming the presence of aggregated structures as observed in the SEM images.

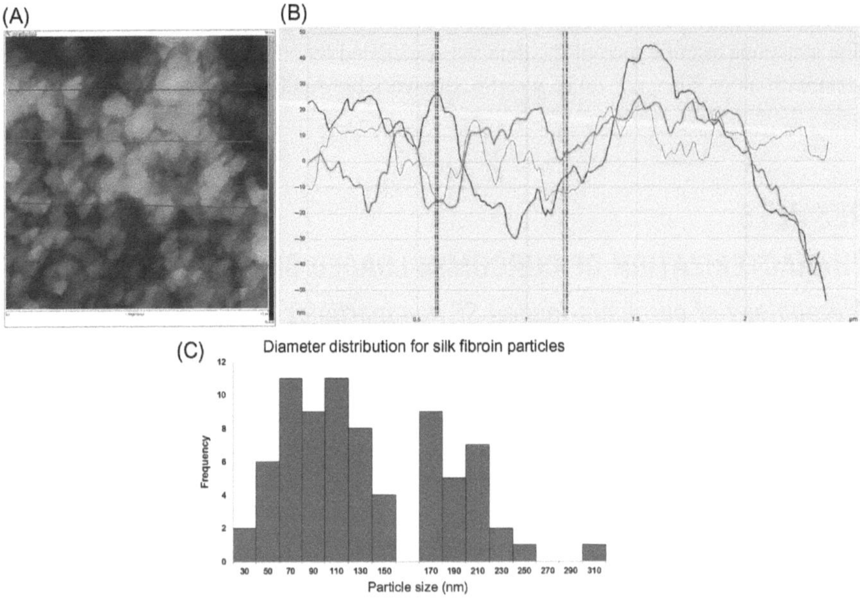

FIGURE 2.6

(A) Image of curcumin-loaded silk nanoparticles obtained using atomic force microscopy. (B) The images were taken using the peak force error channel (left) and height sensor channel (right). The flattened appearance of the nanoparticles results from dehydration of the sample prior to imaging. Average particle diameter was 127.7 ± 6.8 nm (mean ± standard error mean, $n = 78$). (C) Distribution plot of particle size calculated from measurements taken from AFM images. The majority of the particles were found to have diameters in the range of 70–110 nm, $n = 78$.

Drug loading efficiency and release kinetics of curcumin from SF nanoparticles

Drug loading efficiency was calculated to determine the amount of curcumin loaded into the SF nanoparticles. Curcumin loading efficiency was calculated using the equation:

$$\text{Loading efficiency (\%)} = \frac{(CnNP - SFNP)}{(Cn + SFNP)} \times (100\%)$$

where *CnNP* is the weight of the curcumin-loaded silk nanoparticles, *SFNP* is the weight of the silk nanoparticles, and *Cn* is the initial amount of curcumin added. Curcumin content was determined to be 40 μg (of 400 μg added) per 1 mg curcumin-loaded silk nanoparticles, or 4 wt%. The loading efficiency of curcumin into the silk nanoparticles was 10%. Release kinetics of curcumin from the SF nanoparticles was monitored over a 7-day period. Curcumin release from the nanoparticles showed an initial burst release of more than 50% in the first 24 hours, with continued curcumin release up to 7 days.

BIOLOGICAL EVALUATION OF CURCUMIN-LOADED SF NANOPARTICLES

Curcumin efficacy against GILM2 metastatic breast cancer cells

GILM2 metastatic breast cancer cells were treated with free curcumin, SF nanoparticles, or curcumin-loaded SF nanoparticles for 24–96 hours, and cell viability was evaluated using MTT assay. As seen in Fig. 2.7, cell viability decreased linearly as the concentration of curcumin increased, particularly for curcumin concentrations greater than 50 μM. Although it appeared that lower concentrations of curcumin-loaded SF nanoparticles were less toxic after 96 hours of treatment compared to 24 hours of treatment, prolonged exposure to higher concentrations of curcumin-loaded SF nanoparticles caused significant decreases in cell viability.

Protein array analysis of apoptotic pathways activated by curcumin

The protein core facility returned an excel sheet containing data regarding protein expression levels for each of the 171 proteins arrayed. A heat map was generated

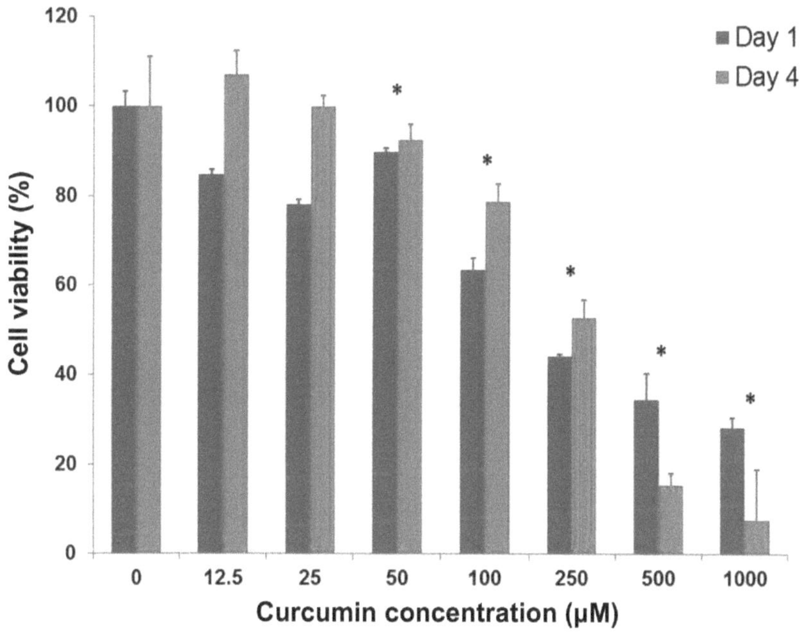

FIGURE 2.7

Cell viability following treatment with curcumin-loaded silk fibroin nanoparticles in GILM2 metastatic breast cancer cells. Cell viability was measured using MTT assay. Cell viability directly correlated with the dose of curcumin; as the dose of curcumin increased, a direct decrease in cell viability was observed, indicating that higher curcumin concentrations induced greater toxicity in this cell type. The exact dosage of curcumin delivered to GILM2 cells was calculated based on the curcumin content in the nanoparticles. $n = 9$, * indicates significance of $p < .05$ relative to control (no curcumin treatment).

to graphically represent expression patterns across the samples. The data were normalized against the control condition and entered into IPA software (Ingenuity Systems, Redwood City, CA). As shown in Fig. 2.8, 171 different proteins were evaluated using reverse phase protein microarray, which are sorted according to their location in the cell. Of those 171 proteins, curcumin treatment directly or

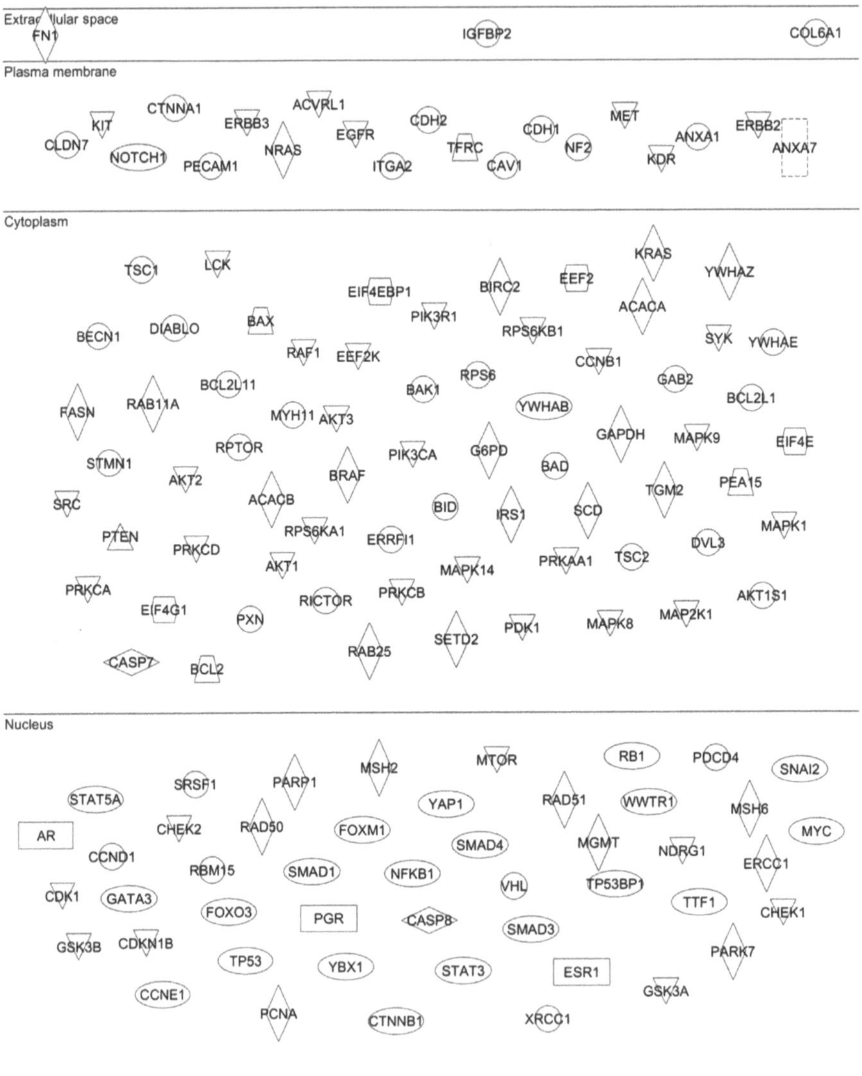

FIGURE 2.8

A complete list of the 171 proteins probed for the protein microarray analysis. Proteins are sorted according to their location in the cell.

Table 2.2 Subset of Proteins Directly Affected by Curcumin Treatment

AR	BIRC2	CDKN1B	FASN	MTOR
BAX	CASP7	CTNNB1	GSK3B	MYC
BCL2	CASP8	DIABLO	IRS1	NFkB1
BCL2L1	CCNB1	EGFR	MAPK1	NOTCH1
BCL2L11	CDH1	EIF4E	MAPK8	
BID	CDK1	ERBB2	MAPK9	

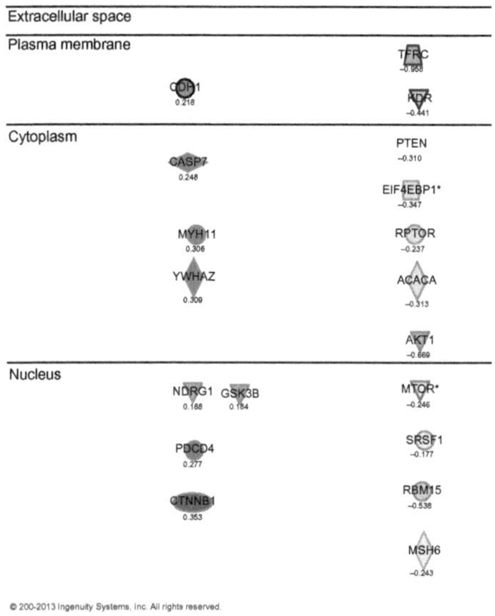

Upregulated proteins	Downregulated proteins
CDH1	TFRC
CASP7	KDR
MYH11	PTEN
YWHAZ	EIF4EBP1
NDRG1	RPTOR
GSK3B	ACACA
PDCD4	AKT1
CTNNB1	mTOR
	SRSF1
	RBM15
	MSH6

FIGURE 2.9

Subset of molecules whose expression is regulated directly or indirectly by free curcumin.

indirectly affected the expression of a subset of 29 proteins, which are listed in Table 2.2. Free curcumin treatment caused the upregulation of 8 proteins and the downregulation of 11 proteins, shown in Fig. 2.9. Treatment with curcumin-loaded nanoparticles caused the upregulation of 18 proteins and the downregulation of 13 proteins, shown in Fig. 2.10. The proteins whose expression was affected by free curcumin treatment were different from the proteins affected by the curcumin-loaded nanoparticles treatment, indicating that encapsulation of the curcumin within the silk nanoparticles changed how the curcumin interacted with the GILM2 cells.

Basal protein expression profiles in untreated control GILM2 cells are shown in Fig. 2.11. Ras expression led to high expression of AKT, particularly high expression of AKT1. The downstream effects of this upregulation led to cell cycle regulation and oncogenesis, promoted cell survival, suppression of apoptosis and

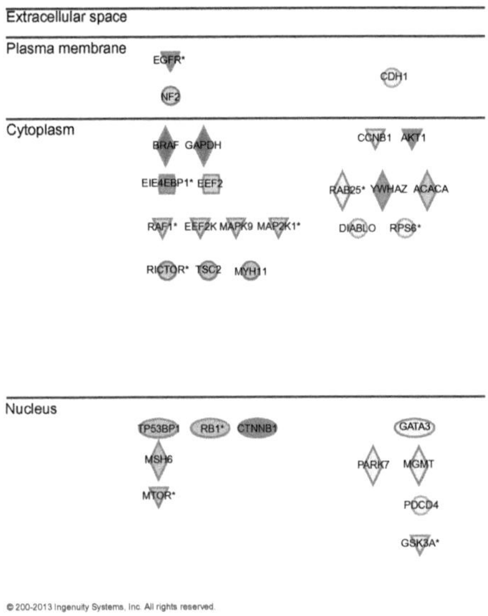

Upregulated proteins	Downregulated proteins
EGFR	CDH1
NF2	CCNB1
BRAF	AKT1
GAPDH	RAB25
EIF4EBP1	YWHAZ
EEF2	ACACA
RAF1	DIABLO
EEF2K	RPS6
MAPK9	GATA3
MAP2K1	PARK7
RICTOR	MGMT
TSC2	PDCD4
MYH11	GSK3A
TP53BP1	
RB1	
CTNNB1	
MSH6	
mTOR	

FIGURE 2.10

Subset of molecules whose expression is regulated directly or indirectly by curcumin-loaded silk nanoparticles.

cell cycle arrest. Notably, we saw expression of the tumor suppressor protein p53 in GILM2 cells although cells were not undergoing apoptosis, suggesting that this could be a mutant form of p53.

Based on the IPA analysis, protein expression for the cell group treated with 50 μM free curcumin for 24 hours was correlated with existing canonical pathways, and a list of more than 200 potential pathways were reported sorted by the highest concurrence rate, of which the top 5 pathways are listed in Fig. 2.12. The top three pathways were selected for further analysis: PI3K/AKT signaling, insulin receptor signaling, and ILK signaling.

Within the PI3K/AKT signaling pathway, known to affect cell proliferation, there was significant downregulation of AKT1, which led to downstream downregulation of mTOR, which caused suppression of protein synthesis and also affected cell growth. Additionally, downregulation of AKT1 also promoted the upregulation of GSK3B, leading to significant upregulation of β-catenin and affecting the WNT pathway (Fig. 2.13).

Within the insulin receptor signaling pathway, downregulation of AKT1 caused a decrease in expression of the downstream molecule TSC2, inhibiting its effect on mTOR signaling. mTOR was also affected by downregulation of Raptor and 4E-BP1 which collectively affected protein synthesis (Fig. 2.14).

(A)

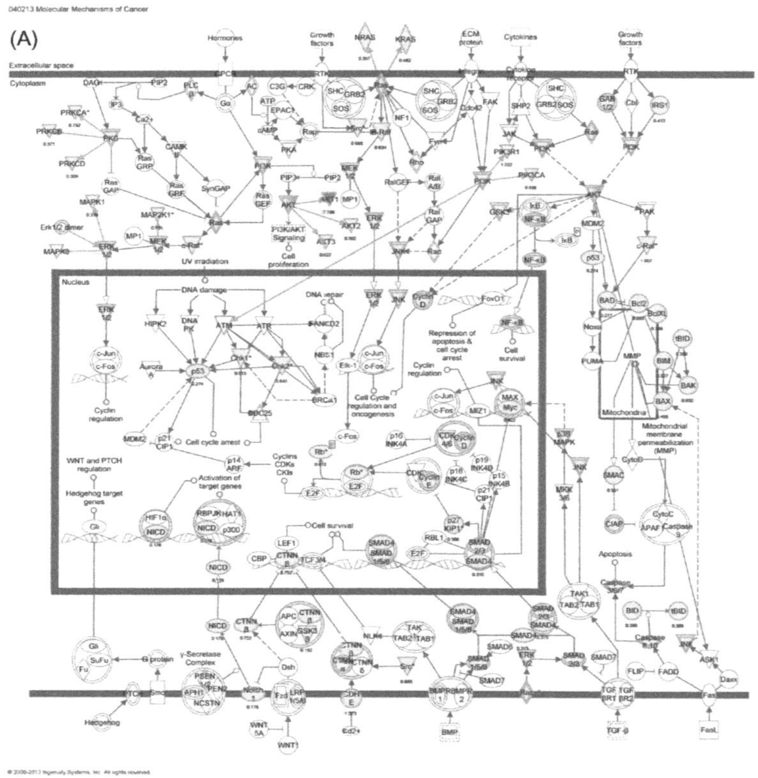

(B)

High expression (>0.75)	Moderate expression (0.5–0.74)	Low expression (0.3–0.49)	Slight expression (<0.3)
AKT1	B-Raf	AKT2	AKT3
BAK	Chk1	BAX	BAD
c-Raf	Chk2	BclXL	Bcl2
CDHE	PIK3CA	BIM	GSK3β
CTNNβ	Rb	IRS1	MAPK1
MAP2K1	SMAC	KRAS	NICD
PIK3R1	SRC	Myc	p53
PRKCA		NRAS	
		p27KIP1	
		PRKCB	
		PRKCD	
		SMAD4	
		tBID	

FIGURE 2.11

(A and B) Protein expression profile in control untreated GILM2 cells.

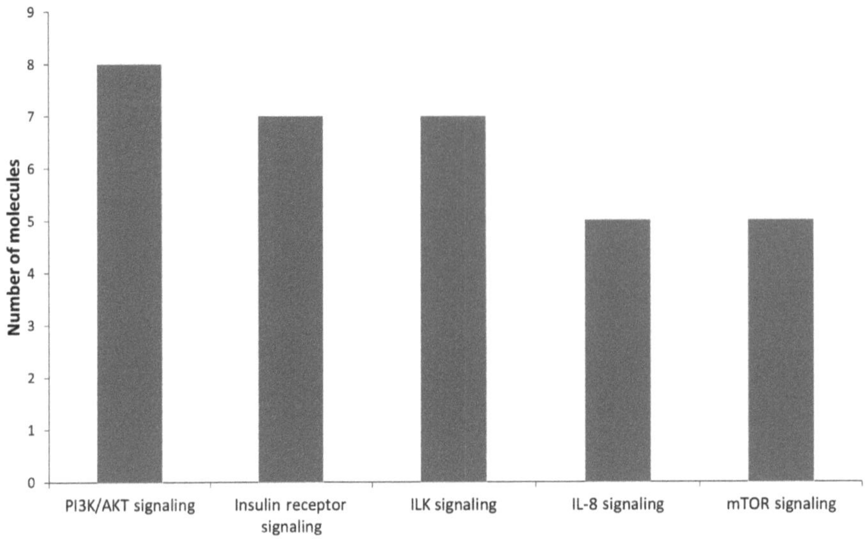

FIGURE 2.12

Canonical pathways identified from protein expression profiles following treatment with 50 μM free curcumin for 24 h.

The ILK pathway is known to affect cell migration, cell proliferation, cell adhesion, and signal transduction. Free curcumin treatment caused the upregulation of β-catenin and GSK3B, although it did not change the expression levels of c-Myc and cyclin-D1. Therefore the upregulation of GSK3B prevented the transport of β-catenin from the cytoplasm to the nucleus, thereby preventing cell proliferation (Fig. 2.15).

Based on the IPA analysis, protein expression for the cell group treated with 50 μM curcumin-loaded nanoparticles for 48 hours was correlated with existing canonical pathways, and a list of more than 200 potential pathways were reported sorted by the highest concurrence rate, of which the top 5 pathways are listed in Fig. 2.16. The top three pathways were selected for further analysis: molecular mechanisms of cancer, p70S6K signaling, and PI3K/AKT signaling.

Within the molecular mechanisms of cancer, there was significant downregulation of AKT1 and upregulation of B-Raf, which both affected cell cycle regulation and oncogenesis. In addition, downregulation of AKT1 affected cell proliferation through the PI3K/AKT pathway. Upregulation of Rb expression promoted cell cycle arrest, while upregulation of CTNNβ expression suppressed cell survival. Slight upregulation of c-Raf and MAP2K1 expression affected cyclin regulation (Fig. 2.17).

The p70S6K signaling pathway plays a role in protein synthesis and in cell growth control. Within this pathway, significant downregulation of AKT1 expression led to the upregulation of mTOR expression. Upregulation of EGFR caused upregulation of c-Raf and MAP2K1, which also contributed to upregulation of

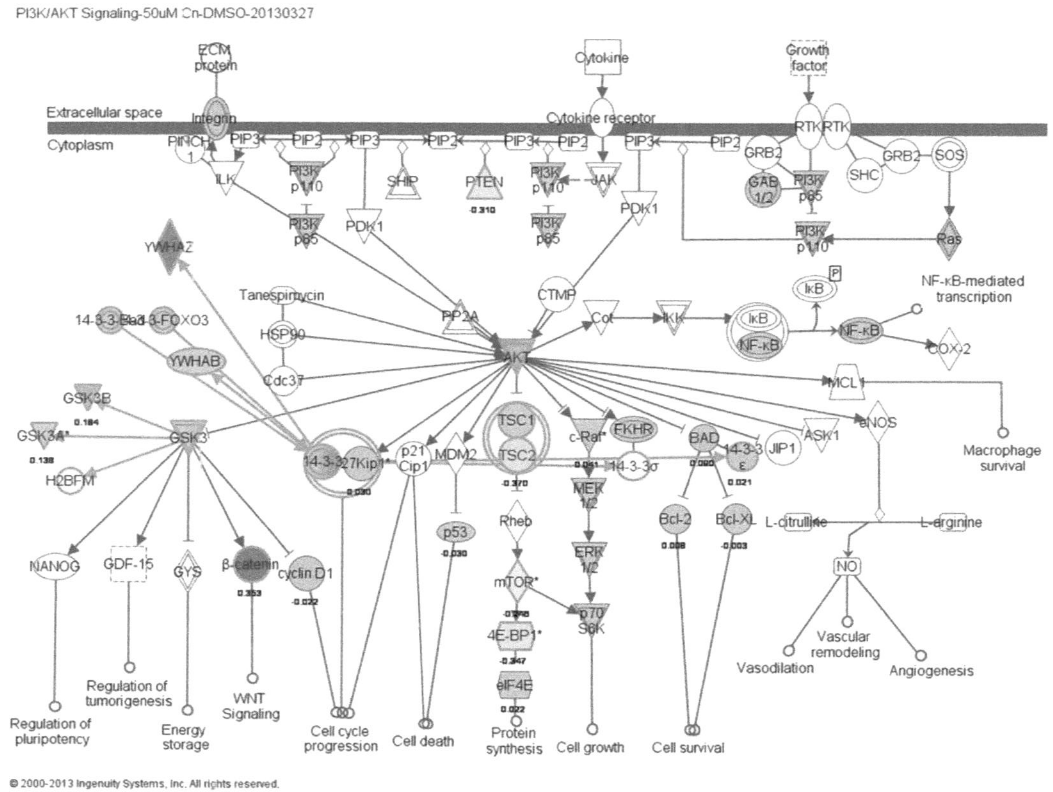

FIGURE 2.13

PI3K/AKT signaling pathway—protein expression profiles following 50 μM free curcumin treatment for 24 h.

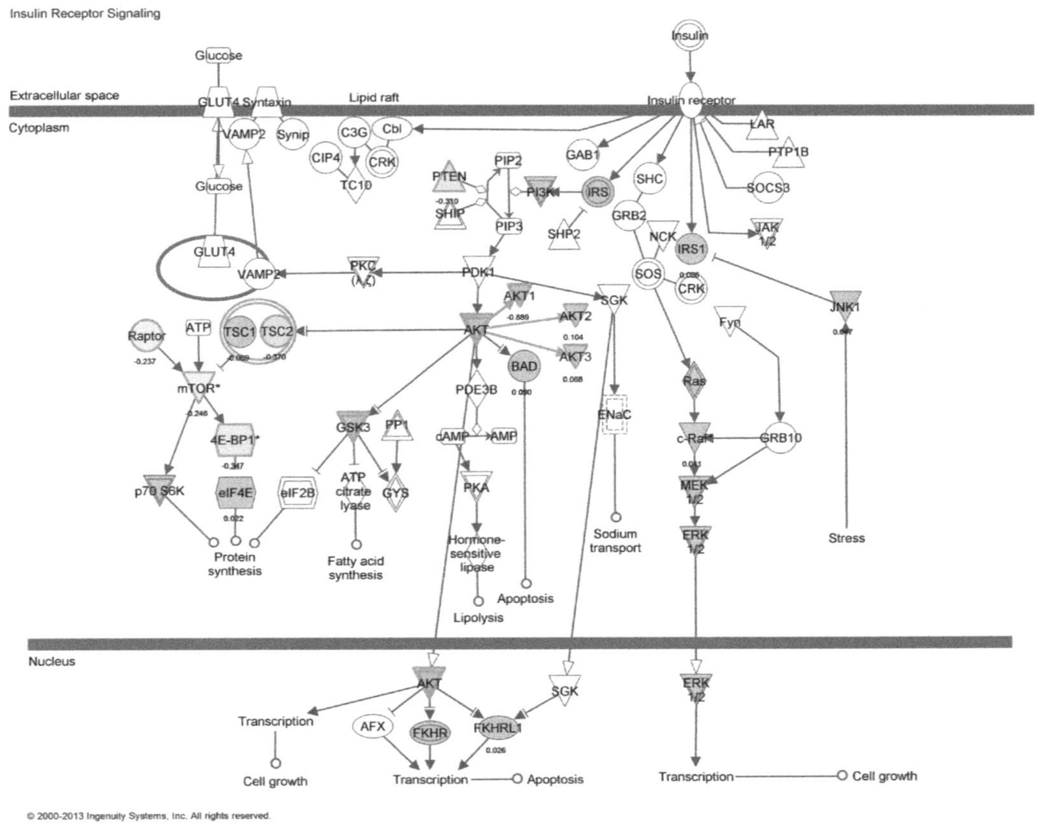

FIGURE 2.14

Insulin receptor signaling pathway—protein expression profiles following 50 mM free curcumin treatment for 24 h.

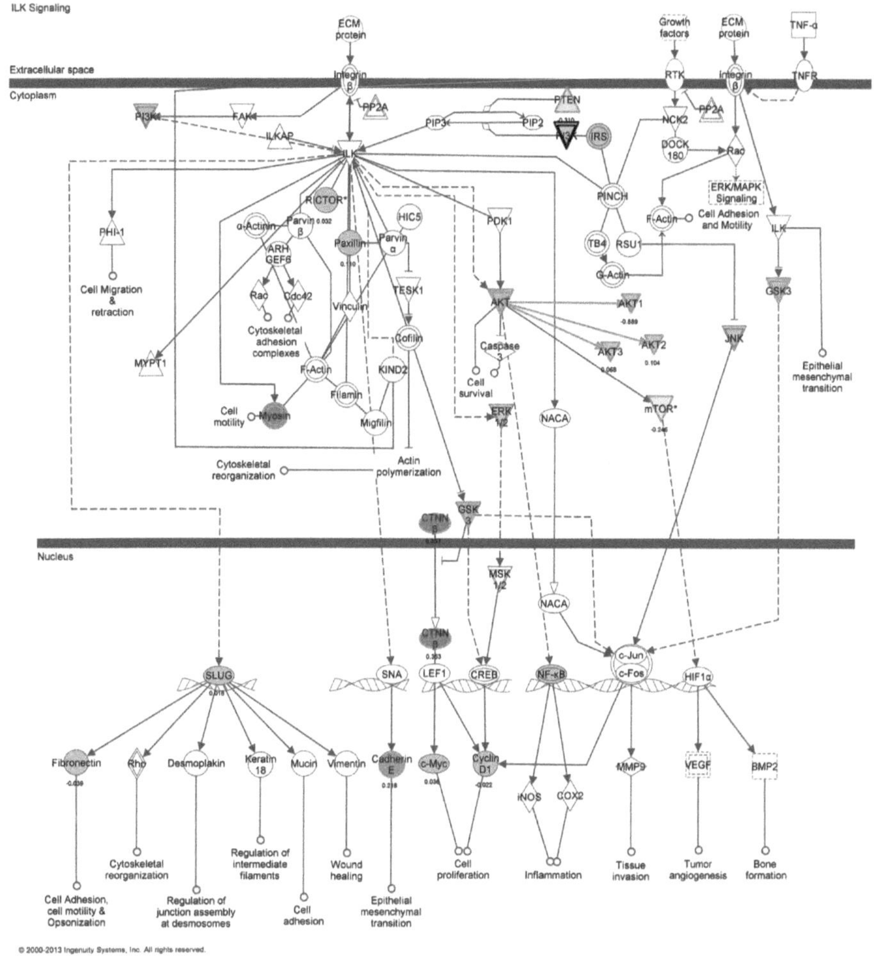

FIGURE 2.15

ILK signaling pathway—protein expression profiles following 50 μM free curcumin treatment for 24 h.

mTOR. This in turn decreased expression of S6 and increased expression of EEF2K which inhibited EEF, which had the downstream effect of suppressing protein synthesis. Upregulation of mTOR also caused the downregulation of 14-3-3 which suppressed cell survival (Fig. 2.18).

Within the PI3K/AKT signaling pathway, known to affect cell proliferation, the downregulation of AKT1 triggered the upregulation of c-Raf and MAP2K1 which affected cell growth, and triggered the upregulation of mTOR and 4E-BP1 which caused inhibition of protein synthesis. Downregulation of AKT1 also led to the downregulation of 14-3-3 which suppressed cell cycle progression. Finally,

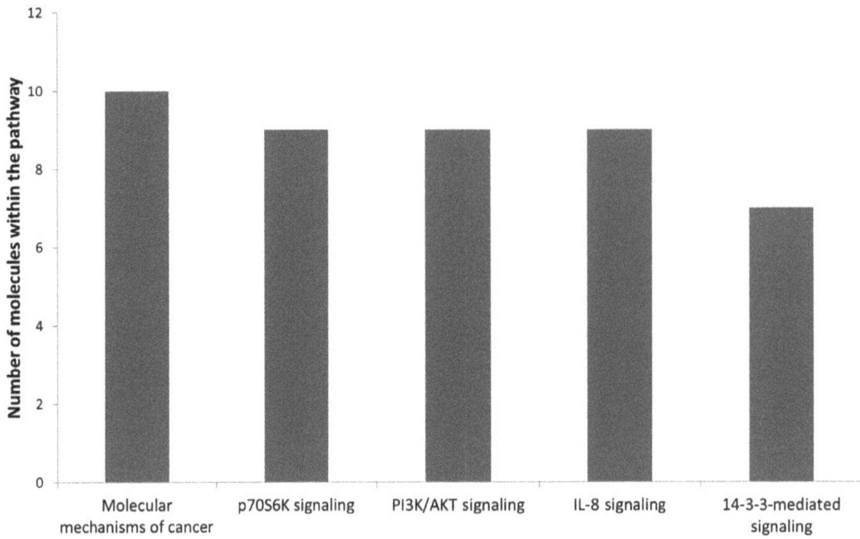

FIGURE 2.16

Canonical pathways identified from protein expression profiles following treatment with 50 μM curcumin-loaded nanoparticles for 48 h.

downregulation of AKT1 caused the downregulation of GSK3A which upregulated β-catenin expression and affected the WNT signaling pathway as mentioned previously for free curcumin treatment (Fig. 2.19).

DISCUSSION

While great strides have been made in the treatment and understanding of breast cancer, less progress has been made in improving treatment options for patients who are diagnosed with or develop metastatic breast cancer. The most common treatment options are typically less effective, as metastatic cancer cells have evolved sophisticated methods of drug resistance. Side effects of these treatments tend to decrease patients' quality of life, so using potent agents derived from nature would decrease side effects without compromising efficacy. Delivering chemotherapy agents to the target location in higher amounts would reduce the total dose administered, thereby also decreasing side effects. Unfortunately current drug delivery carriers do not accumulate efficiently in tumor tissues. Nanoparticle-based systems have been shown to increase accumulation in the tumor while protecting the drug payload during circulation. Additionally, using biodegradable materials such as SF eliminates concerns about toxicity due to

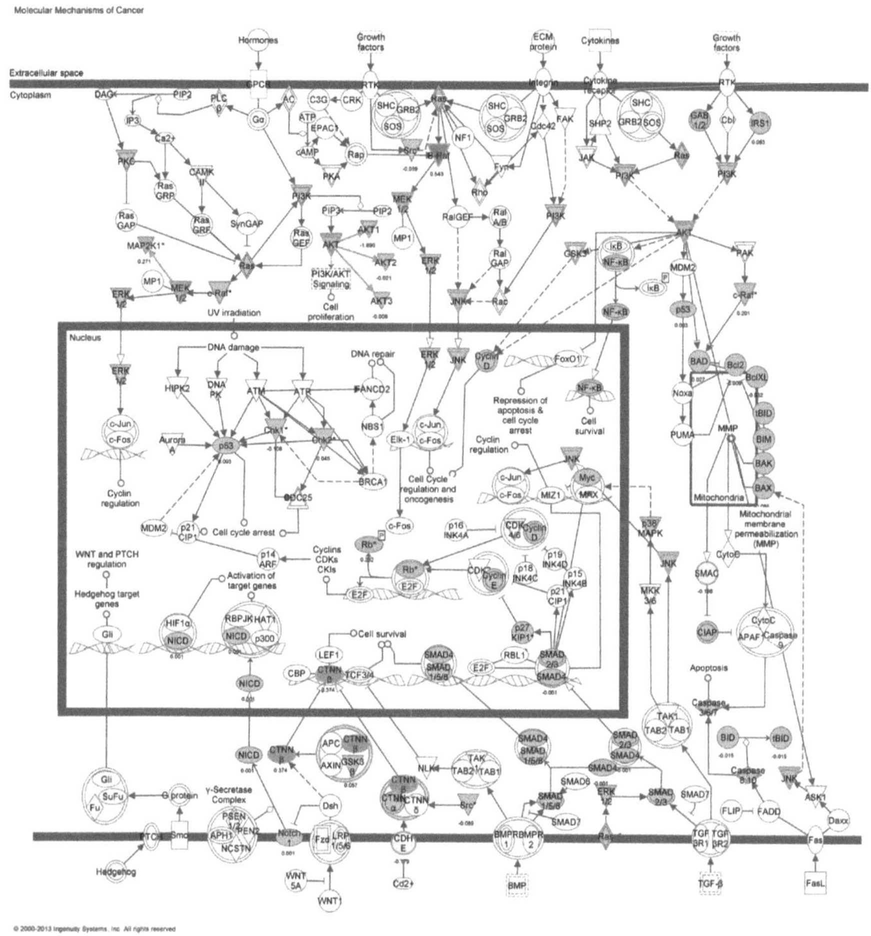

FIGURE 2.17

Molecular mechanisms of cancer—protein expression profiles following 50-μM curcumin nanoparticle treatment for 48 h.

material accumulation. Our goal in this study was twofold: to determine if curcumin prevent GILM2 metastatic breast cancer cell survival, and the underlying molecular mechanisms causing cell death.

Despite the promises curcumin has shown as a potential antiproliferative and anticancer agent in vitro, its hydrophobicity prevents significant bioavailability upon administration. Preclinical evaluation of curcumin through oral, intraperitoneal, or intravenous administration routes has shown that curcumin has very low bioavailability; therefore delivery and significant drug accumulation at target tumor sites requires a stable drug delivery system.

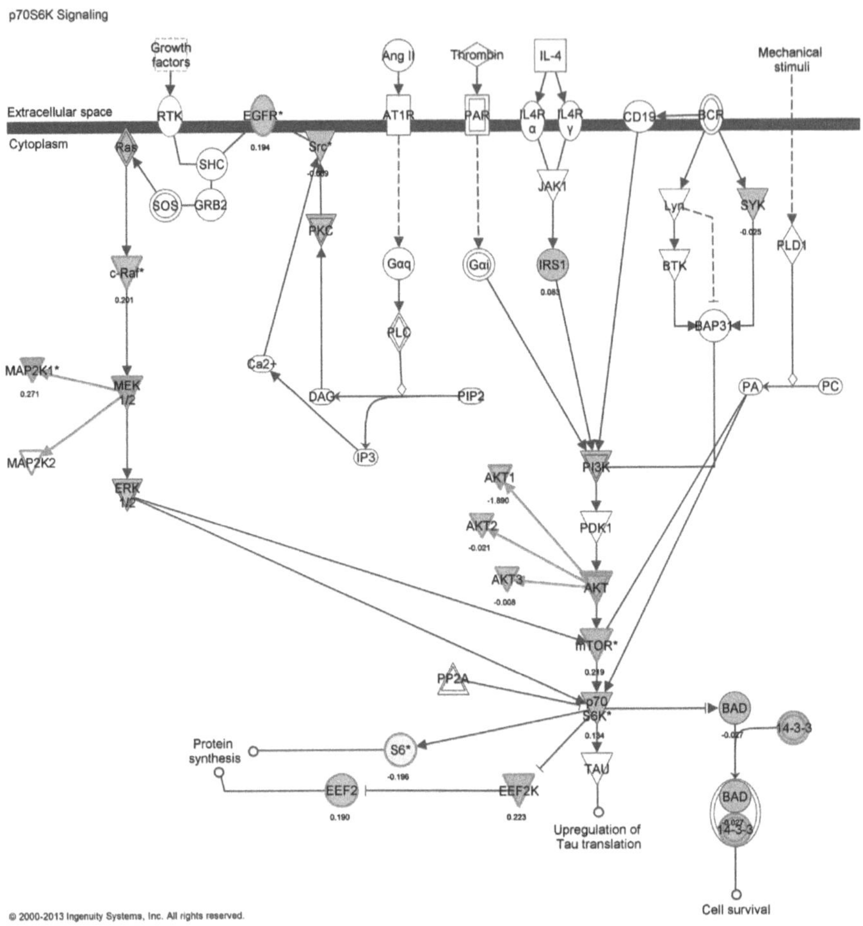

FIGURE 2.18

p70S6K Signaling pathway—protein expression profiles following 50-μM curcumin nanoparticle treatment for 48 h.

Proteins such as SF have been shown to contain side chain hydrophobic methyl groups which can bind highly hydrophobic drugs such as curcumin. The unique structure of SF allows the self-assembly of beta barrel structures where hydrophobic groups form an inner core surrounding curcumin while the hydrophilic groups are located on the outer surface interacting with the physiological milieu [24]. Our initial studies demonstrated the efficacy of two different SF particle systems in treating various breast cancer cell lines. One study reported the fabrication, characterization, and efficacy of curcumin-loaded silk nanoparticles [25], while a

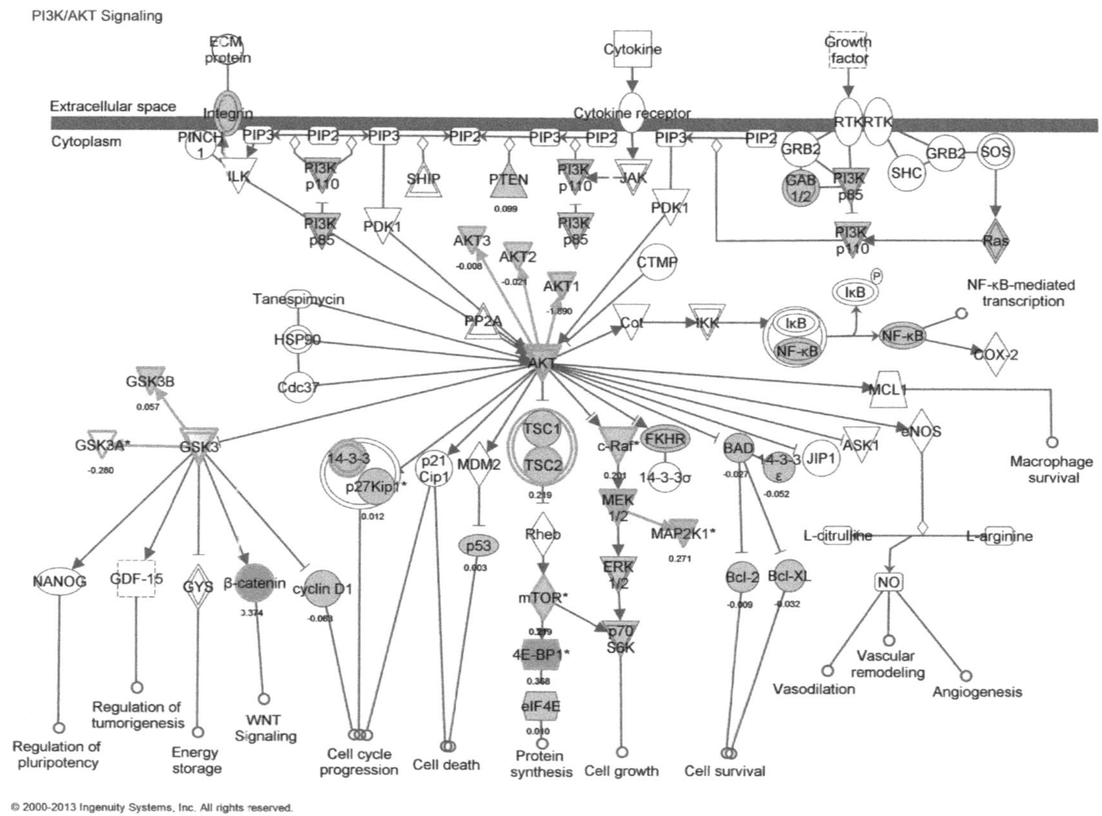

FIGURE 2.19

PI3K/AKT signaling pathway—protein expression profiles following 50-μM curcumin nanoparticle treatment for 48 h.

second study evaluated silk fibroin-coated emodin-loaded liposomes [24]. Our lab has also demonstrated that SF could be used to fabricate drug (emodin) releasing scaffolds which could be implanted posttumor resection during flap reconstruction to prevent tumor recurrence. The structure of SF also allows the potential for further chemical modifications of the nanoparticle surface to add tumor specific ligands which would improve targeting and increase drug payload delivered to specific tissue/tumor sites.

In this study, we used a novel preparation method to create curcumin-loaded SF nanoparticles, whereby we used solvent precipitation to form the nanoparticles, but dissolved the drug (curcumin) in the solvent (ethanol) prior to nanoparticle formation. Stable particle diameters of around 130 nm were detected by AFM, although in PBS solution nanoparticles aggregated into large clusters about 600 nm in diameter as detected by DLS. Moreover, electron microscopy images of the particles showed clustered structures of 130-nm particles suggesting that the particles indeed existed in aggregated form when suspended in aqueous solution.

Our cell viability data showed that curcumin appeared has more efficacies at lower concentrations of curcumin-loaded nanoparticles during the initial 24 hours of treatment, while viability increased for those same concentrations after 96 hours. This could be due to the fact that the amount of curcumin released was not cytotoxic to all of the cells, and therefore cells were able to recover and proliferate within the 96 hours incubation time, leading to an increase in overall cell viability at 96 hours. Cytotoxicity associated with curcumin was significant for 50 μM and higher concentrations at both 24 and 96 hours.

Protein microarray analysis allowed us to rapidly screen changes in protein expression simultaneously for 171 different proteins. IPA software allowed us to input the protein expression data to identify and correlate groups of proteins with existing canonical pathways. Basal protein expression in untreated GILM2 cells showed that these cells had evolved mechanisms to overcome p53 mediated apoptosis, thereby suppressing apoptosis and cell cycle arrest and promoting cell survival and oncogenesis. Following treatment with free curcumin for 24 hours, there were dramatic changes in protein expression, particularly in the downregulation of AKT1 expression. This downregulation caused a myriad of downstream events including the suppression of protein synthesis, cell proliferation, and cell survival, collectively leading to cell death. The tumor suppressive gene PTEN was upregulated in untreated control GILM2 cells, while curcumin treatment downregulated this molecule, indicating a reversal of tumor promotion which would lead to cell death, and possibly decrease tumor size in vivo. Downregulation of AKT1 and mTOR could also prevent HIF1 transcription of VEGF within the cell, thereby preventing tumor angiogenesis in vivo which could lead to decreased tumor size. These results indicate that treatment with curcumin could prevent tumor growth, which could lead to more favorable patient outcomes in the clinic.

CONCLUSIONS

We demonstrated a novel method of preparing curcumin-loaded silk fibroin nanoparticles based on solvent precipitation. These curcumin-loaded silk fibroin nanoparticles were efficacious in decreasing cell viability in vitro in GILM2 metastatic breast cancer cells. Pathway analysis indicated that curcumin caused significant downregulation of AKT1 expression, which had the downstream effects of suppressing protein synthesis, cell survival, and oncogenesis, therefore leading to cell death. Future studies will evaluate the efficacy of this nanoparticle system in vivo to confirm the effect of curcumin in a living animal, which would complete preclinical evaluation of this curcumin-loaded silk nanoparticle drug delivery system.

ACKNOWLEDGMENTS

This research was supported by Kyte Family Gift Fund, Longenbaugh foundation, TRAMCEL Program Provost funds, and National Institutes of Health NIHR0134658. We also thank the High Resolution Electron Microscopy Facility (HREMF; Cancer Center Core Grant CA16672).

REFERENCES

[1] Desantis C, Siegel R, Bandi P, Jemal A. Breast cancer statistics. CA Cancer J Clin 2011;61:409–18.

[2] Howlader N, Noone AM, Krapcho M, Neyman N, Aminou R, Altekruse SF, et al. SEER cancer statistics review, 1975–2009 (Vintage 2009 Populations). Bethesda, MD: National Cancer Institute; 2012.

[3] O'shaughnessy J. Extending survival with chemotherapy in metastatic breast cancer. Oncologist 2005;10:20–9.

[4] Largillier R, Ferrero JM, Doyen JEA. Prognostic factors in 1,038 women with metastatic breast cancer. Ann Oncol 2008;19:2012–19.

[5] Cardoso F, Harbeck N, Fallowfield L, Kyriakides S, Senkus E, on behalf of the ESMO Guidelines Working Group. Locally recurrent or metastatic breast cancer: ESMO clinical recommendations for diagnosis, treatment and follow-up. Ann Oncol 2012;20:15–18.

[6] Aggarwal BB, Sundaram C, Malani N, Ichikawa H. Curcumin: the Indian solid gold. Adv Exp Med Biol 2007;595:75.

[7] Zbarsky V, Datla KP, Parkar S, Rai DK, Aruoma OI, Dexter DT. Neuroprotective properties of the natural phenolic antioxidants curcumin and naringenin but not quercetin and fisetin in a 6-OHDA model of Parkinson's disease. Free Radic Res 2005;39:1119–25.

[8] Calabrese V, Butter DA, Stella AM. Nutritional antioxidants and the heme oxygenase pathway of stress tolerance: novel targets for neuroprotection in Alzheimer's disease. Ital J Biochem 2003;52:177–81.
[9] Hatcher H, Lanalp R, Cho J, Torti FM, Torti SV. Curcumin: from ancient medicine to current clinical trials. Cell Mol Life Sci 2008;65:1631–52.
[10] Aggarwal BB, Kumar A, Bharti AC. Anticancer potential of curcumin: preclinical and clinical studies. Anticancer Res 2003;23:363–98.
[11] Epstein J, Sanderson IR, Macdonald TT. Curcumin as a therapeutic agent: the evidence from in vitro, animal and human studies. Br J Nutr 2010;103(11):1545–57.
[12] Anand P, Kunnumakkara AB, Newman RA, Aggarwal BB. Bioavailability of curcumin: problems and promises. Mol Pharm 2007;4:807–18.
[13] Greco F, Vicent MJ. Polymer–drug conjugates: current status and future trends. Front Biosci 2008;13:2744–56.
[14] Jain RK, Stylianopoulos T. Delivering nanomedicine to solid tumors. Nat Rev Clin Oncol 2010;7:653–64.
[15] Klok HA. Peptide/protein-synthetic polymer conjugates; quo vadis. Macromolecules 2009;42:7990.
[16] Lee CC, Mackay JA, Frechet JMJ, Szoka FC. Designing dendrimers for biological applications. Nat Biotechnol 2005;23:1517.
[17] Malam Y, Loizidou M, Seifalian AM. Liposomes and nanoparticles: nanosized vehicles for drug delivery in cancer. Trends Pharmacol Sci 2009;30:592–9.
[18] Allen TM, Cullis PR. Drug delivery systems: entering the mainstream. Science 2004;303:1818–22.
[19] Christian DA, Tian AW, Ellenbroek WG, Levental I, Rajagopal K, Janmey PA, et al. Spotted vesicles, striped micelles and Janus assemblies induced by ligand binding. Nat Mater 2009;8:843–9.
[20] Davis ME, Chen Z, Shin DM. Nanoparticle therapeutics: an emerging treatment modality for cancer. Nat Rev Drug Discov 2008;7:771–82.
[21] Zhang L, Gu FX, Chan JM, Wang AZ, Langer RS, Farokhzad OC. Nanoparticles in medicine: therapeutic applications and developments. Clin Pharmacol Ther 2008;83:761–9.
[22] Harries M, Ellis P, Harper P. Nanoparticle albumin-bound paclitaxel for metastatic breast cancer. J Clin Oncol 2005;23:7768–71.
[23] Gobin AS, Froude VE, Mathur AB. Structural and mechanical characteristics of silk fibroin and chitosan blend scaffolds for tissue regeneration. J Biomed Mater Res 2005;74A:465–73.
[24] Gobin AS, Rhea R, Newman RA, Mathur AB. Silk-fibroin coated liposomes for long-term and targeted drug delivery. Int J Nanomed 2005;1:81–7.
[25] Gupta V, Aseh A, Rios CN, Aggarwal BB, Mathur AB. Fabrication and characterization of silk fibroin-derived curcumin nanoparticles for cancer therapy. Int J Nanomed 2009;4:115–22.

CHAPTER

Animal models in cancer nanotechnology

3

S.L. Craig[1] and V.B. Jensen[2]

[1]*Medical University of South Carolina, Charleston, SC, United States* [2]*The University of Texas MD Anderson Cancer Center, Houston, TX, United States*

CHAPTER OUTLINE

Nanoparticles are synthetic constructs, 1–100 nm in diameter, made of organic and inorganic materials. Nanoparticles have been engineered to detect cancer, to treat cancer (by targeting specific tissues, reaching subcellular compartments, or targeting malignant cells in the circulation), and to monitor the progress of various cancer treatments [1].

Animals are used in research to achieve scientific understanding of the whole body system. They provide a valuable tool with which to study disease processes and methods of preventing or treating those disease processes. Cancer scientists use animals for a multitude of research projects. Animal models of cancer are used to determine the pharmacokinetics of nanoparticles, to diagnose cancer using imaging modalities, and to treat cancer using nanotechnology. Nanomaterials can carry small-molecule drugs, small interfering RNA (siRNA), or vehicles for imaging and thermal absorption [1]. Here we present several animal models that illustrate the use of nanoparticles in cancer.

NANOPARTICLES' USE IN PHARMACOKINETICS

Nanoparticles can be used to track the distribution of the nanoparticles themselves and any agent that they may be carrying after injection in animal models of cancer. This has been done in various ways using ex vivo methods, imaging, and

Nanotechnology in Cancer.

measurement of blood levels. Polyethylene glycol (PEG)-encapsulated (PEGylated) liposomes are nanoparticles that improve drug delivery to tumors but do not damage normal tissue. PEGylated encapsulations can carry drugs such as doxorubicin, liposomal belotecan (an analog of camptothecin), or cisplatin. The amount of drug available is determined by calculating the pharmacokinetics. Pharmacokinetics can vary between species, and there is variability between human patients due to the mononuclear phagocyte system, which releases the drug from the liposome package by digesting it. Non-PEGylated nanoparticles release a drug more rapidly than PEGylated nanoparticles do. The pharmacokinetics was measured in various species including mice, rats, dogs, and humans with refractory solid tumors [2].

Measurement of drugs in the bloodstream when the drug is administered with nanoparticles in an animal model of cancer can help determine the biodistribution of the drug along with changes when delivered with nanoparticles. It has been noted that many multidrug-resistant cancers overexpress epidermal growth factor receptor (EGFR). Fourteen athymic nude mice were injected directly into the mammary fat pad with hypoxic MDA-MB-231 breast cancer tumor cells in Matrigel. These mice were the animal model for a study involving EGFR-targeted and EGFR-nontargeted polymer-blend nanoparticles. These PEGylated nanoparticles were loaded with paclitaxel and lonidamine. Paclitaxel is commonly used to treat cancer but has known toxic effects; lonidamine inhibits aerobic glycolysis and induces apoptosis but it, too, has toxic effects (liver toxicity). The combination of these drugs in clinical trials was shown to improve the efficacy of both drugs in a synergistic manner. The goal of this study was to measure levels of both drugs in the tumor and the bloodstream of tumor-bearing mice. Both the targeted and nontargeted nanoparticles had increased biodistribution when compared with the drug solution which was a combination of the drugs (paclitaxel and lonidamine) without nanoparticles. Targeted nanoparticles did improve the pharmacokinetics of lonidamine [3].

Nanoprobes have been used to determine their targeted distribution in a tumor model. Angiogenesis is a critical event enabling tumor growth. Vascular endothelial growth factor (VEGF) and receptor (VEGFR) signaling pathways play an essential and rate-limiting role in promoting tumor-induced angiogenesis. Not only does angiogenesis correlate with the onset of tumor development but also with growth, metastasis, and invasion of tumors. A 100-nm nanoprobe encapsulated with an iodine contrast agent was used to determine tumor vessel permeability using digital mammography [4].

A female Fischer rat was subcutaneously implanted with cells from the 13762 MAT B III rat mammary adenocarcinoma cell line. On Day 7 after inoculation, intravenous injection of a nanoprobe that encapsulated an iodine contrast agent was imaged with the use of clinical digital mammography. The deposition of the nanoprobes usually coincided with regions of high levels of VEGFR-2, which indicate leakier blood vessels. Thus there is enhanced accumulation of nanoprobes in regions of high angiogenic activity [4].

An ex vivo biodistribution model was utilized to track where nanoparticles accumulated in targeted tissue. Lipid-based oil-filled nanoparticles with chelated nickel targeted to EGFR-overexpressing epidermoid carcinoma cells (A431) were injected into nude mice bearing A431 tumors. EGFR is a transmembrane tyrosine kinase receptor that regulates cell proliferation, apoptosis, differentiation, and migration. EGFR is overexpressed in many cancers including cancers of the breast, prostate, ovary, bladder, pancreas, lung, and kidney, as well as glioma, to name a few. In this model, affinity proteins were used to target the tumor. Affinity proteins, composed of 58 amino acid residues bundled into a three-helix scaffold, are small and easy to produce and can have affinity for specific receptors such as EGFR. After the nanoparticles were injected into the nude mouse xenograft model, the mice were euthanized. Sixteen hours after intravenous injection of the nanoparticles, 19% were detected in tumor tissue, 28% in the liver, and 42% in the kidneys. This finding can be used as a tool for targeting drug delivery to EGFR-positive cancers [5].

Biodistribution of nanoparticles can be tracked using various routes of delivery. YIGSR is a pentapeptide laminin-binding site often found on metastatic cancer cells. In the Sarfati et al. study (2011), YIGSR covalently attached to fluorescent nanoparticles (YIGSR-NPs) was injected subretinally into a CB57B/6 mouse model of melanoma. The mouse melanoma model was created by injecting the mice either subretinally or subcutaneously with cells from the M-cherry-labeled B16F10.9 melanoma cell line CRL-6326. The purpose of the subcutaneous injections of tumor cells was to verify that the YIGSR-NPs honed to the tumor; the purpose of the subretinal injection of tumor cells was to duplicate lung metastasis [6]. Mice that had the solid tumor development after subcutaneous injection of melanoma were given YIGSR-NPs intravenously and the targeted nanoparticles honed to the tumor. The control solid tumor group had a greater disposition of nanoparticles in the liver and spleen. Mice with solid melanoma tumors were injected intratumorally with YIGSR-NPs and they were able to retain the nanoparticles after 48 hours. The control group cleared 75% of the nanoparticles from the tumor in 48 hours that were not covalently attached to YIGSR. Mice injected subretinally with the melanoma cells established the lung metastasis and micro- and macro-metastases within 12 days of injection. Following subretinal injection of the YIGSR-NPs, there was a two- to threefold accumulation of nanoparticles in the lungs at 6 and 24 hours compared to controls where accumulation of the noncovalently bound nanoparticles accumulated in the liver and spleen. In addition, this study also suggested that the nanoparticles were excreted through the urine [6].

Quantum dots (QDs) have been used as a marker to track the movement of a single nanoparticle using a dorsal skinfold mouse model. Female BALB/c athymic nude mice were implanted subcutaneously with cells from the human breast cancer cell line KPL-4, which overexpresses HER2. When the tumor reached a volume of 100–200 mm^3, a surgical dorsal skin flap was created in the mice by using two sterilized polyvinyl chloride plates containing a window. The movement of the QDs was observed in the dorsal skinfold chamber using a confocal microscope and camera. This technique allowed the observation of a single nanoparticle in circulation, at extravasation, extracellularly, bound to HER2 on the cell membrane, and moving into the perinuclear region. Ultimately the processes of delivery of QD antibody constructs were identified and analyzed and the rate-limiting parameters could be looked at in vivo [7].

NANOPARTICLES AS DIAGNOSTIC IMAGING TOOLS

Accurate imaging tools are important in staging, treatment, and prognosis of cancer. Nanoparticles have value as being useful in diagnostic imaging using radioisotopes, combined with fluorescent dyes or antibodies, or having the ability to be active in a magnetic field for magnetic resonance imaging (MRI). Sentinel lymph nodes are used when staging breast cancer and melanoma. The regional lymph nodes can help with prognosis by helping predict the metastatic status of a solid organ tumor. Current methods of detection use lymphoscintigraphy, which involves injecting radiolabeled nanoparticles throughout the tumor followed by nuclear imaging or detection with a gamma probe to track the lymphatic drainage. The most common gamma-emitting tracer used to label nanoparticles is technetium-99. These labeled nanoparticles are called radiocolloids. Metastases in the sentinel lymph nodes indicate a poor prognosis. Another method for detecting sentinel lymph nodes is to use blue dyes and trace the drainage. The combined use of radiocolloids and blue dye improves the detection rate.

There are other optical lymphotropic imaging agents such as the combination of indocyanine green with nanoparticles that fluoresce and are captured using near-infrared (NIR) cameras. In addition, photoacoustic imaging is another optical modality using optically active contrast agents such as gold nanorods or carbon nanotubes used in combination with a pulsed laser and ultrasound.

Another technique used for imaging in the animal model is computed tomography (CT) by lymphotropic tracers. One of the issues discovered was that the amount of gold nanoparticles that are needed would be large, based on pig and mouse studies [8–12]. Mouse models have been used to trace the path of the lymph system as it relates to the primary tumor. In one mouse model the nanoparticulate lymphotropic contrast agent of gold nanoparticles conjugated to anti-CD45 antibodies was injected into the toe of either the front or back paw of a mouse. In this model the contrast agent had a high affinity for the popliteal or the axillary lymph node [9,13].

Nanoablation is a therapy that intensifies the uptake of superparamagnetic iron oxide (SPIO) nanoparticles in both hepatic and nonhepatic tumors compared with standard intravenous dosing. Intratumoral SPIO nanoparticle uptake after nanoablation can be successfully quantified noninvasively with 7-T MRI. Imaging can be used as a method to estimate localized drug delivery after nanoablation. SPIO nanoparticles act as MRI contrast agents because their superparamagnetic core causes more rapid T1 and T2 relaxation of the immediately surrounding tissues [14].

This technique has been applied to Sprague Dawley rats using the N1-S1 rat hepatoma cell line. The animal model was created via percutaneous ultrasound-guided tumor implantation of the N1-S1 hepatoma into the left lateral lobe of the liver. Nanoablation was performed by exposing the left lateral liver lobe via a laparotomy. In addition the left femoral vein was catheterized for injection of DOX-SPIO nanoparticles. Electroporation was performed for 2 minutes after DOX-SPIO nanoparticle injection via a two-pronged electroporation tool. Upon completion of this process, the liver was returned to the abdominal cavity and the incision closed. The rats were euthanized within 10 minutes after nanoablation and immediately returned to the MRI scanner after treatment. With this technique, MRI can be used to measure the uptake of chemotherapeutic drugs conjugated to nanoparticles [14].

Nanoparticles in conjunction with the iron oxide have been used in another animal model for MRI work. A rat model of metastasis was used to test the use of ultrasmall superparamagnetic iron oxide (USPIO) in combination with MRI to detect sentinel lymph nodes. This model used ACI-AXC 9935/Irish rats injected in the rear paw with cells from the hepatoma H-4-II E tumor cell line, with expected metastasis to the popliteal and paraaortic lymph node within 3 weeks after injection. After metastasis was established, rats were injected with the USPIO preparation intravenously and MRI was performed. In rats that did not have metastasis, signal was dissipated within 24–48 hours after injection, whereas in the metastases model, the signal did not decrease in this same time frame [9,15].

Another model using nanoparticles for imaging involves gold nanoparticles and angiogenesis. Angiogenesis has also been studied in BALB/cBYJNarl mice by using gold nanoparticle colloids. These colloids can be used for other purposes such as cancer targeting, diagnosis, or radiotherapy. To study diffusion of contrast agents through vessel leakage, these mice were injected with the mouse colon carcinoma cell line CT-26 subcutaneously in the thigh. This cell line is known to be highly metastatic. After approximately 7 days, when the tumors reached 100–120 mm^3, and again after approximately 26 days, when the tumors reached 1000–1200 mm^3, the mice were injected through the tail vein with PEG-Au nanoparticles. These gold nanoparticles behave like hydrophilic contrast agents and localize in the tumor after leaking out of the microvessels using enhanced retention and permeation effect which was visualized using microradiology [8].

Imaging using nanotechnology and angiogenesis has utilized lipid nanobubbles as a contrast agent. When tumors induce angiogenesis, the blood vessels are leaky due to large pore cutoff sizes and large fenestrations. These defects in the blood vessels created through angiogenesis allow for delivery of drugs and gene carriers to the tumor tissue, known as the enhanced permeability and retention (EPR) effect. Lipid nanobubbles that were used as a contrast agent for ultrasound were fabricated and evaluated. The nanobubbles were <450 nm. Mouse prostatic RM-1 cancer cells were subcutaneously injected on the dorsal scapular area in BALB/c athymic nude mice and allowed to grow to a diameter of 1.2 cm. Nanobubbles were injected intravenously via the tail vein. The tumor was then imaged via ultrasound for 1.5 hours on anesthetized mice. The nanobubbles improved contrast enhancement for approximately an hour. The red fluorescently dyed nanobubbles were also viewed ex vivo using a confocal laser scanning microscope. The nanobubbles were injected into the tail vein of tumor-bearing mice and after perfusion of the heart with saline, the tumors and muscle of the thigh were removed and viewed with the use of a confocal microscope. The nanobubbles were present in the intracellular spaces in the tumors but were not readily evident in the muscle sections, indicating passive targeting of the tumor [16].

Combining different imaging modalities can increase the sensitivity of diagnosis and perhaps the treatment of cancer. QDs are small semiconductor nanocrystals with a narrow emission spectrum and high intensity of fluorescence; they are often smaller than 100 nm. QDs are typically composed of cadmium selenide (CdSe) [17]. QDs were used recently for in vivo imaging such as sentinel lymph node mapping and embryo development, tumor angiogenesis, and tracking of metastasis. The NIR region of 700–900 nm allows biomolecules to reach a minimum absorbance for in vivo optical imaging. Arginine-glycine-aspartic acid (RGD) peptide-conjugated NIR QDs can be used for tumor vasculature targeting and imaging in living mice. A combination of near-infrared fluorescent (NIRF) imaging, positron emission tomography (PET), and QDs can provide a sensitive and quantitative assessment of the pharmacokinetics and tumor-targeting efficacy of the NIRF QDs. This may lead to the development of fluorescence-guided surgery and other clinical applications. The nude mouse model was used by injecting the mice subcutaneously with U87MG human glioblastoma cells in the flank. The mice were imaged after ^{64}Cu-labeled cell adhesion molecule RGD peptide-conjugated QD was injected intravenously. The mice were then imaged with use of micro-PET in vivo, and harvested tumors were imaged ex vivo. The RGD-QD combination targeted the tumor vasculature with minimal extravasation and this dual-function probe could reduce toxicity and overcome tissue penetration limitations of optical imaging [18]. QDs labeled with RGD peptide sequence have been used to image glioblastoma tumors implanted in mice [19].

Knowing the tumor boundaries can be very useful in cancer treatment. Nanoparticles can assist in better defining the boundary of a solid tumor. Oral squamous cell carcinoma is a common head and neck tumor that is typically treated surgically. The ability to define the boundary of the tumor for surgical removal can improve survival rate and quality of life. Oral squamous carcinoma animal model was developed by injecting female BALB/c nude mice subcutaneously in the cheek with cells from the human buccal squamous cell carcinoma cell line BcaCD885. Oral squamous cell carcinoma tumors highly express EGFR, and this was targeted by EGFR monoclonal antibody–conjugated QDs injected intravenously into the mice. Images were maximal between 15 minutes and 6 hours after injection of the conjugated QDs. Images were captured with use of a confocal microscope [20].

Nanoparticles have been used with confocal microscope technology to enhance the ability to visualize tumor cells. QDs linked to alpha-fetoprotein (AFP) antibody (QDs-Anti-AFP) were used as a marker for hepatocellular carcinoma. QDs may be used as an alternative to organic immunoflourescent probes for cancer detection due to their high sensitivity and specificity for cancer cells. Nude mice were implanted subcutaneously with cells from the HCCLM6 human hepatocellular carcinoma cell line. Tumors were allowed to reach 0.5–1 cm in diameter, and then QDs-Anti-AFP probes were injected intravenously. The mice were euthanized and the tissues were examined with use of a confocal fluorescence microscope. This method allowed active tumor targeting and spectroscopic hepatic carcinoma imaging [21].

Image-guided radiotherapy enhances precise delivery of radiation to tumor tissue. A liquid tissue marker that serves as a fixed area (fiducial) for multiple dosing was used in the following study. This liquid marker was made of nanogel and due to its high radiopacity allows for marker-based image guidance in 2D and 3D X-ray-based imaging during radiotherapy. A canine cancer patient with a spontaneous subcutaneous mast cell tumor was used as a model for evaluating the efficacy of a nanogel marker. The nanogel consisting of an acylated derivative of sucrose, polylactic acid, and ethanol plus acrylamide-coated AuNPs was injected intratumorally while the dog was under general anesthesia. Approximately 24 hours after the nanogel injection, a CT scan of the tumor/nanogel region was performed. This was followed by radiation therapy consisting of 4 fractions of 6 Gy radiation over a 16-day period. Radiation therapy was delivered based on the position of the nanogel. This study concluded the nanogel remained constant in size and volume over the time of the treatment. It continued to retain its 3D shape and remained immobilized throughout the course of treatment despite the canine patient exercising or resting, with no side effects noted. This nanogel represents a liquid fiducial marker with high contrast for 2D X-ray imaging. By injecting the gel into the tumor, the AuNP may enhance the effect of the radiation therapy. Tumor tracking in radiotherapy may improve treatment outcomes through the production of a more precise delivery method [22].

Imaging of the Thomsen–Friedenreich (TF) antigen, which is often overexpressed in colorectal cancer, was performed in the rat model. A nanobeacon was developed that would allow visualization of the TF antigen with use of a FL microendoscope. HCT116 cells were injected into the descending colon of 8- to 10-week-old female athymic nude rats (rnu/rnu, homozygous), after which a white light colonoscopy and then fluorescent endoscopy were performed. This nanobeacon allowed visualization of the tumor and may be used to track tumor regression after treatment through fluorescence colonoscopy [23].

In addition to being a useful therapeutic agent, nanotechnology can be a used as a diagnostic tool. Nanoliposomes can function as carriers of radionuclides for targeting solid tumor. By encapsulating therapeutic radionuclides for internal targeted radiotherapy, nanoliposomes can act as a carrier system. Radiotherapeutics (^{188}Re-liposome) and radiochemotherapeutics (^{188}Re-DXR-liposome) administered intravenously to nude mice bearing subcutaneous human HT-29 colorectal adenocarcinoma xenografts provide not only a tool for signal photon-emission computed tomography (SPECT) imaging but also a drug carrier for treating

HT-29 solid tumor. Micro-SPECT imaging was performed within 48 hours of injection of ^{188}Re-(DXR)-liposomes. The ^{188}Re-DXR-liposome accumulated in the liver, spleen, and tumor. In addition tumor uptake could be clearly seen over time. This formulation provides a useful diagnostic tool and delivery system to the tumor, liver, and spleen [24].

NANOPARTICLES AS A THERANOSTIC TOOL

The combination of diagnostic imaging and therapy is called theranostic. Theranostics has an increasingly important role in personalized cancer medicine. One goal is to improve solubility of functional agents, protect them from premature degradation, prolong blood circulation, and enhance tumor accumulation [25].

The uses of nanoparticles for theranostic purposes have been done in animal models of cancer. PEGylated nanoparticles are often used to prolong the vascular circulation of drugs used to combat cancer. There are concerns, however, associated with administration of PEGylated nanoparticles, namely, immunogenicity, anti-PEG immune response, biocompatibility, and toxic effects associated with chronic administration. PEG-free, porphyrin-based ultrasmall nanoparticles that mimic lipoproteins (PLPs) have been used to integrate both imaging and therapeutic functions. These PLPs are stable in blood circulation and rapidly dissociate once they accumulate in a tumor, allowing for fluorescence imaging and tumor-selective photodynamic reactivity and therapy [25].

Another theranostic nanoparticle is porphysomes, organic optically active nanovesicles formed from porphyrin bilayers. Porphysomes are nontoxic in mice and have highly self-quenched energy. Porphysomes allow the visualization of lymphatic systems with use of photoacoustic tomography. Photothermal therapy utilizes contrast agents that convert light to heat in a targeted manner. When exposed to laser irradiation, the porphysomes' energy is released thermally, similar to what occurs in gold nanorods [26].

Nude mice were implanted subcutaneously with KB tumors that were grown for 2–3 weeks. Then the mice were injected with bacteriochlorophyll porphysomes via the tail vein for fluorescence imaging. Nude mice with KB cell subcutaneous xenografts that were injected intravenously with porphysomes and 24 hours later, they underwent photothermal therapy with use of a laser. The tumor showed a therapeutic response with increased survival time for the mice after the photothermal ablation. Additionally the mice underwent NIR resonance imaging and tumor temperature was measured using an infrared camera. The tumors fluoresced 2 days after injection due to accumulation of the porphysomes in the tumor [26]. Unlike the PLPs, the porphysomes are known to require PEG to increase the stability and avoid rapid clearance by the reticuloendothelial system (RES) [25].

Biocompatible gum Arabic–stabilized gold nanocrystals (GA-AuNPs) were a theranostic agent that was used as an X-ray contrast agent in a tumor-bearing dog. The data suggested that accumulation of GA-AuNPs reaches a threshold limit within a short period (5 hours). The nanocrystals were retained in the tumor tissue for 24 hours [27].

AuNPs have also been used as theranostic tools. A castrated male client-owned dog was presented with a large ventral cervical mass consisting of a mixture of a thyroid carcinoma and osteosarcoma. Metastasis to the lungs was present at the time of diagnosis. With use of a CT scan, intratumoral injections of GA-AuNP were administered under general anesthesia at seven sites of the tumor for a total volume of 2.1 mL. The dog was anesthetized 24 hours after the nanoparticles were injected, and a second CT scan was performed. Images obtained after treatment revealed a slight increase in Hounsefield units (HU) of contrast enhancement of the cervical mass compared with the preinjection CT scan. Increased HU values indicated uptake in the intratumoral mass of the dog. Noted adverse effects were mild edema and mild fever after injection. At 3 weeks after injection, the dog was euthanized due to progressive clinical signs of local disease. This procedure provides a protocol for multiple-injection of intratumoral nanoparticles to ensure effective accumulation of substantial quantities of nanoparticles within a tumor site to produce contrast enhancement [27].

Additionally in a study by Nurunnabi et al. [28], green graphene quantum dots (GQDs) were tested for theranostic abilities in BALB/c athymic nude mice; GQDs were implanted subcutaneously with KB tumor cells (KERATIN-forming tumor cell line HeLa). At 21 days after tumor injection, in vivo imaging was performed. Although photoluminescent GQDs had accumulated in the liver, heart, spleen, lung, kidneys, and tumor sites after intravenous injection of the QDs, the fluorescence signal was observed only in the tumor site with use of a molecular imaging system. The GQDs were approximately 5 nm in diameter [28].

NANOPARTICLES AS A TREATMENT TOOL

One of the major goals of cancer therapeutics is to kill cancer cells while not damaging normal cells. One way to achieve this is the use of molecularly targeted therapy combined with chemotherapy. Tissue and cell distribution of cancer therapeutic drugs can be controlled by entrapment in submicronic (<1 μm) colloidal systems, otherwise known as nanoparticles. Nanoparticles have been shown to potentially reverse multiple-drug resistance. Some of the desirable characteristics that are needed to deliver therapeutic agents to tumor cells include the ability to overcome drug resistance at the tumor and cellular levels and to ensure appropriate distribution, biotransformation, and clearance of the drug [29].

To investigate nanoparticles for treatment purposes, the nanoparticle should be directed to the location of the tumor. First nonmodified nanoparticles were looked at to see where they were deposited after intravenous injection. Nonmodified nanoparticles in the bloodstream are cleared by fixed macrophages typically found in the liver, spleen, lungs, and bone marrow. This was verified by injecting mice in the tail vein with [^{14}C]-doxorubicin incorporated into polyisohexylcyanoacrylate nanoparticles. This nanoparticle combination had a higher concentration of doxorubicin in the liver, spleen, and lungs compared with concentrations in mice given free doxorubicin, which resulted in higher levels in the heart and kidneys [29,30]. In the metastases model of C57BL/6 mice injected intravenously with reticulosarcoma cell line M 5076, doxorubicin-loaded polyisohexylcyanoacrylate nanoparticles increased the antitumoral cytotoxic activity in the liver [31].

Similar to the use of polyisohexlcyanoacrylate as a carrier for doxorubicin, nanoparticles can provide yet another carrier-mediated drug targeting system for tumors. To counter the fast removal of particles from the circulation by the RES, nanosized particles combined with a hydrophilic surface can delay RES uptake. CD1-Nu mice were implanted subcutaneously with cells from the pancreatic human cell line PANC-1. Once the tumor size reached 75 mm^3, nude mice were injected intravenously with doxorubicin (DXR)-loaded targeted hyaluronan liposomes, tHA-LIP-DXR on Days 0, 7, and 14. By Day 32, the response of the animals treated with tHA-LIP-DXR indicated tumor volumes significantly smaller than that at initiation of treatment. Moreover, some mice were tumor-free [32].

Liposomal delivery systems provide modified techniques for improved pharmacokinetic and safety profiles of cytotoxic drugs. The antitumor activity of a nanoliposomal formulation of irinotecan (nal-IRI) can achieve greater intratumoral levels of the prodrug irinotecan and its active metabolite SN-38 compared with free irinotecan. Human colorectal adenocarcinoma HT-29 cells were injected subcutaneously into the right flank of NOD/SCID mice. After the desired tumor volume was achieved, treatments with control, free irinotecan, and nal-IRI were injected intravenously weekly for 4 weeks. Thus the extended exposure of tumor cells to SN-38, which is achieved by nal-IRI, can contribute toward enhanced cytotoxicity compared with free irinotecan. Liposomal encapsulation of irinotecan can safely improve its antitumor activity in preclinical models by enhancing accumulation of its active metabolite within the tumor microenvironment [33].

Nanoliposomal irinotecan has been used as treatment in orthotopic brain tumor models. Chemotherapy has proven to be particularly challenging due to the blood–brain barrier (BBB) and its associated low permeability to a wide variety of drugs. Irinotecan, a widely used cancer chemotherapeutic, displays antitumor activity against various cancer types, including malignant gliomas. Irinotecan has a complex metabolic profile that may limit its ability to provide adequate concentration at the tumor site. Irinotecan is mainly a prodrug, dependent upon conversion to SN-38 by carboxylesterases for optimal anticancer activity [34].

One example of the use of nanoliposomal irinotecan was in athymic nude rats which were injected intracranially with cells from the U87MG glioblastoma multiforme cell line. At 5 days after tumor cell implantation, nanoliposomal irinotecan was administered intravenously. At the end of the study, at Day 100 after tumor implantation, the surviving animals were euthanized and necropsied. The results indicated that nanoliposomal irinotecan was associated with prolonged tumor accumulation/retention and increased survival duration. Furthermore, nanoliposomal delivery of irinotecan produced very high concentrations and prolonged exposure of irinotecan within brain tumors, which was consequently associated with increased intratumoral SN-38 levels. Nanoliposomal irinotecan accumulates progressively to high levels within brain tumors by utilizing the EPR effect of angiogenesis, followed by intratumoral conversion to SN-38. These findings suggest that nanoliposomal irinotecan administered systemically provides significant pharmacologic advantages and may be an efficacious therapy for brain tumors [34].

Various liposomal packaging systems have been developed as primary carriers to avoid phagocytosis and to circulate longer in the blood. An example is folate receptor (FR)-targeted liposomal oridonin, otherwise known as FR-targeted liposomal ORI or F-L-ORI. Tumor-bearing mice were developed by inoculating a suspension of human hepatocellular carcinoma HepG2 cells subcutaneously in athymic nude BALB/c mice. Five days after inoculation, intravenous F-L-ORI was administered. The results indicated that F-L-ORI had produced an antitumor effect in vivo against HepG2 cells [35].

Some tumors are known to have a high number of folic acid receptors. Female BALB/c mice were injected in the footpad with cells from the murine lung carcinoma cell line, M109R-HiFR. This cell line is known to express folic acid at a high level. FR-targeted liposomes that were loaded with doxorubicin were exposed to the M109R-HiFR cells in vitro and then injected into the mice. These cells exhibited less tumor growth than did untreated cells [36,37]. Similarly, FR-targeted lipid nanoparticles loaded with lipophilic paclitaxel was injected into tumor-bearing mice. Paclitaxel is typically formulated in Cremophor EL, which has several adverse effects including nephrotoxicity and hypersensitivity. Female BALB/c mice were injected subcutaneously into the flank with M109 cells and then injected intraperitoneally with lipid FR-targeted nanoparticles with paclitaxel. The tumor growth rate for the mice that received the targeted nanoparticles was less than that for the mice that received paclitaxel in Cremophor EL [38].

There are various nanoliposomal formulations that are effective in treating cancer. Nanoscale ceramide liposomes can be loaded with a variety of anticancer compounds to create a combination therapy against tumors. C6-ceramide is a sphingolipid metabolite that causes cancer cell death. When C6-ceramide is encapsulated in a nanoliposome bilayer formulation, cell death is selectively induced. To create the animal model, BALB/c athymic nude mice were irradiated (600 cGy) 1 day before inoculation with cells from the chronic lymphocytic leukemia (CLL) cell line, JVM3. Cells were injected subcutaneously into the right flank of the mice. Treatment began approximately 2 weeks after inoculation. Mice were then treated with C6-ceramide nanoliposomes intravenously, which preferentially inhibit the altered metabolism of glucose in leukemic cells via downregulation of glyceraldehyde-3-phosphate dehydrogenase (GAPDH), resulting in induction of necrotic cell death [39].

Bilateral human hepatocellular carcinoma tumor xenografts were established in athymic nude mice by subcutaneous injection of SK-HEP-1 cells. One week after tumor cell inoculation, C6-ceramide nanoliposomes were injected intravenously. Administration of nanoliposomal C6-ceramide restricted angiogenesis and induced tumor cell apoptosis, thereby preventing the growth of human hepatocellular SK-HEP-1 tumor xenografts in athymic nude mice [40].

Nanoliposomal C6-ceramide in combination with vinblastine has been used to enhance cell death through apoptosis. Nanoliposomal C6-ceramide is an autophagy inducer, and vinblastine is an autophagy maturation inhibitor. Autophagy is the recycling of cellular proteins and organelles during periods of starvation and is used for survival of the cell. Autophagy is also used to remove damaged organelles and long-lived protein. During solid tumor development, the cancer cells are rapidly multiplying and need nutrients to survive. Autophagy during this phase with poor vascularization allows the solid tumor during the progressive phase to survive. To study this mechanism, athymic nude mice were injected subcutaneously in the flank with cells from the human colon adenocarcinoma cell line LS174T. The nanoliposomal C6-ceramide was injected intravenously via the tail vein followed by vinblastine 15 minutes later. This combination treatment resulted in suppression of tumor growth when compared to nanoliposomal C6-ceramide and vinblastine alone [41].

Another animal model for the use of ceramide for treatment of leukemia using nanoliposomes was the Fischer 344 rat. Ceramide as noted previously has been recognized as an antiproliferative and proapoptotic sphingolipid metabolite. Cells from the RNK-16 in vivo T-cell large granular lymphocyte (LGL) leukemic cell line were intraperitoneally transplanted into Fischer 344 rats. Five weeks after inoculation, leukemic rats were treated with C6-ceramide nanoliposomes via tail vein injections for 6 weeks. The C6-ceramide nanoliposome–treated group exhibited greater survivability than the untreated and ghost nanoliposome (no ceramide) group. More importantly, some of the rats treated with nanoliposomal C6-ceramide had maintained normal blood cell counts without circulating immature blood cells, suggesting achievement of complete clinical remission. The responsive rats were euthanized and necropsied at various time points to further evaluate remission. At necropsy, there was resolution of organomegaly, representing a 3- to 10-fold reduction in weight. In addition, these rats had normal levels of LGL cells in the blood, marrow, lymph nodes, and lung as well as normal splenic histology. This study demonstrated that the nanoliposomal delivery of C6-ceramide to rats with LGL leukemia induces significant apoptotic cell death, which may potentially lead to resolution of leukemic cell infiltration [42].

Nanoliposomal ceramide also has useful properties as an effective antipancreatic cancer therapeutic in combination with gemcitabine or an inhibitor of ceramide neutralization. The use of D-threo-1-phenyl-2-decanoylamino-3-morpholino-1-propanol (PDMP), a glucosylceramide synthase inhibitor, and gemcitabine, a nucleoside analog, enhances the antitumor activity of C6-ceramide liposome. Cells from the human pancreatic cell line PANC-1 were injected subcutaneously in athymic nude mice followed by intravenous injection of nanoliposomal ceramide with gemcitabine. This combination therapy inhibited growth of the tumor [43].

Nanoliposomal delivery systems have been created by attaching tissue-specific antibodies. Nanoliposomes conjugated to thyroid-stimulating hormone (TSH) bind to the TSH receptor (TSHr) on the surface of thyrocytes. The tumor model to study the ability of nanoliposomes to target the thyroid was developed by injecting follicular thyroid carcinoma (FTC-133 cells) subcutaneously into NOD/SCID mice. In this way, the in vivo anticancer efficacy of gemcitabine, gemcitabine-loaded nanoliposomes, and gemcitabine-loaded TSH-nanoliposomes could be evaluated. Free gemcitabine and both formulations containing gemcitabine nanoliposomes were administered intravenously after the subcutaneous tumor reached the desired volume. Gemcitabine-loaded TSH-nanoliposomes elicited the greatest reduction in tumor growth [44].

Radioactive nanoliposomes are not only beneficial for imaging but have therapeutic applicability as well. The efficacy of radioactive rhenium (^{188}Re)-labeled nanoliposomes was also studied in a C26 colonic peritoneal carcinomatosis mouse model. The use of ^{188}Re-liposomes for passively targeted tumor therapy had greater therapeutic effect than did the chemotherapeutic drug 5-fluorouracil (5-FU) in a colonic peritoneal carcinomatosis mouse model. Nanotargeted ^{188}Re-liposomes administered intravenously have been used for internal radiotherapy of peritoneal carcinomatosis and ascites. BALB/C mice were inoculated intraperitoneally with the C26 murine colon carcinoma cell line. Seven days after tumor cell inoculation, colon peritoneal metastatic carcinoma-bearing mice received an intravenous injection of ^{188}Re-liposomes, 5-FU, or normal saline. Results demonstrated better therapeutic efficacy by inhibiting the progression of peritoneal tumor growth for mice that received intravenously administered ^{188}Re-liposomes. In addition, a therapeutic efficacy study exhibited a greater survival time in mice treated with the ^{188}Re-liposomes. Furthermore, Micro-SPECT/CT images showed high uptake and targeting of ^{188}Re-liposomes in ascites and in the tumor, liver, and feces in the colon. This provides a useful tool for in vivo and ascites retention in peritoneal carcinomatosis mice. The information suggests that ^{188}Re-liposomes have potential benefits for imaging along with the treatment of tumor nodules [45].

Nanoliposomes have a myriad of possible compositions making them a flexible solution to deliver drugs to specific targeted tissues such as tumors. Antitumor efficacy of both nontargeted and HER2-targeted nanoliposome formulations of topotecan was studied in prostate and breast tumor xenografts. Nanoliposomal topotecan (nLs-TPT) that was stabilized with use of sucrose octasulfate was shown to have significant antitumor activity against a DU145 human prostate tumor xenograft model. Nude mice implanted with DU145 prostate cancer cells subcutaneously in the upper back were given six doses of nLs-TPT beginning at 6 days (mean tumor volume of 150 mm^3) after tumor implantation; nLs-TPT was administered every 4 days via the tail vein. The results indicated a substantial regression and even complete cures in 5 of 10 mice in the nLs-TPT group. However, weight loss was noted if the number of doses was increased from 4 to 6. Lastly, mice implanted with cells from the human breast cancer cell line BT474 that overexpressed HER2 did not have complete regression of tumor and experienced significant weight loss during drug therapy beyond 5 doses [46].

Docetaxel formulations have been developed for the purpose of improving its solubility. Docetaxel is not readily water-soluble and is often dissolved in polysorbate 80, which is known to cause hypersensitivity, febrile neutropenia, nail changes, and other adverse effects. Docetaxel is a chemotherapeutic drug similar to paclitaxel used to treat various cancers such as cancers of the breast, stomach, head and neck, and prostate, as well as nonsmall cell lung cancer. Docetaxel-loaded PEG copolymer micelles and docetaxel-loaded PEG copolymer thermosensitive hydrogel were used to treat tumors in the mouse model. Female BALB/c mice were injected subcutaneously with 4T1 mouse breast cancer cells, and tumors were allowed to reach 20–100 mm^3 before treatment was initiated. The docetaxel micelle was delivered intravenously every third day, and the docetaxel hydrogel was injected directly into the tumor. The docetaxel micelles had a better outcome on survivability than the docetaxel hydrogel group did [47].

Paclitaxel is a known hydrophobic drug, therefore making it insoluble in water. Typically paclitaxel is solubilized in a combination of ethanol and a polyethoxylated castor oil called Cremophor. Cremophor is known to cause allergic reactions and patients are often given antihistamines and corticosteroids to prevent allergic reactions prior to dosing with paclitaxel [48].

Unmodified paclitaxel, noncovalently mixed with PEGylated hydrophilic carbon clusters (PEG-HCCs), was used as a drug delivery system in mice with orthotopic head and neck tumors. Hydrophilic carbon clusters (HCCs), which are nanoparticles, were covalently modified with polyethylene glycol (PEG-HCCs) and shown to be nontoxic and an effective drug delivery vehicle in nude mice with OSC-19 oral squamous cell carcinoma in the tongue. Paclitaxel mixed with PEG-HCCs (PTX/PEG-HCCs) that were <40 nm was administered intravenously for 12 days after injection of the OSC-19 cells. The tumors were measured twice a week. After 50 days, the tumors treated with the PTX/PEG-HCC significantly decreased tumor volume and increased survival compared to negative controls [48].

Nanotechnology may be used for a vast array of experiments including assisting with the transport of drugs across BBB. Rats have contributed to science as models for various tumor types. Combining tumor therapy with nanotechnology has led to many distinct new approaches for cancer therapy. Polysorbate 80-coated poly(butyl cyanoacrylate) nanoparticles were shown to enable the transport of a number of drugs including the antitumor antibiotic doxorubicin (DOX) across the BBB to the brain after intravenous administration to considerably reduce the growth of brain tumors in rats [49].

Malignancies of the central nervous system are resistant to systemic chemotherapy, primarily due to the presence of the BBB, which limits penetration of antineoplastic drugs into brain tumors. The BBB acts as an anatomical and physiological barrier formed by a monolayer of endothelial cells that exhibit specific properties such as intercellular tight junctions, which prevent paracellular transport [49].

To create an intracranial glioblastoma model, glioblastoma homogenated brain tissue was stereotaxically injected into the right lateral ventricle of a Wistar rat brain. DOX bound to polysorbate 80-coated (butyl cyanoacrylate) nanoparticles was injected intravenously for the delivery of drugs to solid tumors. Administration of nanoparticulate DOX was associated with a considerably lower mortality and weight loss when compared with use of the free drug. It could therefore be concluded that DOX toxicity could be decreased by using nanoparticles [49].

The use of nanoliposomes extends beyond simple drug delivery. Cationic nanoliposomes offer a useful tool as a carrier because of their ability to form complexes with negatively charged siRNA. siRNA technology has great potential as a therapeutic modality for targeted gene silencing in cancer treatment [50]. In the Matters et al. study (2009), athymic nude mice were injected subcutaneously into the flank with cells from the human pancreatic cell line, AsPC-1. At 14 days after inoculation of tumor cells, cationic nanoliposomes were used as a delivery vehicle to administer gastrin siRNA intravenously. Results indicated a 25% reduction in tumor size. Because the half-life of siRNAs in the peripheral blood is short, cationic nanoliposomes are a promising vehicle of choice [51].

Cationic nanolipoplexes can function as pulmonary cellular delivery system for siRNA. In the mouse model of metastatic lung cancer induced by the B16F10 skin melanoma cell line injected intravenously, intratracheal administration of the cationic nanolipoplex delivering myeloid cell leukemia sequence 1-specific siRNA (siMcl1) reduced expression in B16F10 cell lines. Reduced formation of melanoma tumor nodules was observed in the lung, demonstrating the utility of cationic nanolipoplexes for pulmonary delivery of therapeutic siRNA for the treatment of lung cancers and possibly for other respiratory diseases. Intratracheal administration of siMcl1 in ECL nanolipoplexes inhibited the growth of B16F10 cells in lung tissues. ECL nanolipoplexes provide a means for delivering protein-specific siRNAs to lung cells in an aerosolized manner that target malignant proteins [52].

The use of lipid nanoparticles with either mRNA or siRNA allows the delivery of nucleic acid to tumors in a programmable manner. Overexpression of tumor suppressor microRNAs (mRNAs) or inhibition of oncogenic mRNAs may have a therapeutic effect for various cancers. Viral delivery systems have been used to carry mRNA and siRNA. Viral delivery systems may have adverse effects associated with toxicity, immunity, or insertions into the genome. siRNAs interfere with the expression of specific genes, whereas mRNAs are RNA fragments that prevent the production of a protein by binding and destroying the messenger RNA that would have produced the protein. The KP mouse model of autochthonous lung adenocarcinoma has shown upregulation of the oncogenic *KRAS* (Kirsten rat sarcoma) allele and inactivation of the tumor suppressor gene p53. In the KP mouse, tumors were initiated by intranasal inhalation of Adeno-Cre to delete p53 and to activate *KRAS*. Ten weeks after lung tumor initiation, lipid nanoparticles with mRNA or siRNA were injected intravenously into the mice and significantly delayed tumor progression [53].

Point mutations in the *KRAS* oncogene commonly occur in human patients with pancreatic cancer. Micelles, which are nanoparticles, were used to encapsulate siRNAs that target the *KRAS* gene. Female BALB/c athymic nude mice were subcutaneously injected in the flank with cells from the PANC-1 human pancreatic cancer cell line. The nanoparticles were injected into mice intravenously via the tail vein after the tumors reached approximately 100 mm^3. After 21 days of treatment with nanoparticles, tumors were significantly smaller than were tumors in the controls [54].

Gold nanoparticles have emerged as a form of functional nanomaterials for the treatment of cancer. A liposome-gold nanoparticle (LiposAU NP) was synthesized that could be cleared through the hepatobiliary and renal route. This agent is also used for photothermal therapy, which causes cell death associated with damaged DNA. Pharmacokinetics of the LiposAU NP was measured in Swiss albino mice. BALB/c nude mice were implanted with cells from the human fibrosarcoma cell line HT1080-*fluc2-turbo*FP. Mice given the combination of LiposAU NP and laser therapy had a complete regression of tumor [55].

The premise behind the use of gold nanoparticles in photothermal therapy is the fact that they can absorb light and be used to heat and ablate tumors. To ensure that nontargeted cells are not damaged during NIR irradiation, the AuNPs are infrared-transparent until they accumulate in the tumor cells, thereby reducing background heating in the blood and increasing specificity. Cells from the A431 squamous carcinoma cell line were injected subcutaneously into the thighs of female athymic nude mice; once the tumors reached 100–150 mm^3, the mice received an intravenous injection of AuNPs. The treatment, which included NIR irradiation, resulted in complete tumor ablation with no normal tissue damage. In mice that had been given intratumoral AuNP followed by NIR irradiation, there was 89% tumor ablation [56]. Another use of AuNPs in the mouse model was in combination with radiotherapy. Balb/C mice were injected subcutaneously with syngeneic EMT-6 mammary carcinoma. When the tumors reached 50–90 mm^3, the mice were injected intravenously with AuNPs combined with X-ray irradiation, which resulted in eradication of most tumors [57].

Another use of gold nanoparticles as thermal agents is the use of laser-activated nano-thermolysis as a cell-elimination technology (LANTCET). LANTCET is a type of nanobased technique to obliterate tumors. LANTCET locates and damages tumor cells by generating intracellular photothermal bubbles around clusters of gold nanoparticles [58].

A demonstration of LANTCET is as described. A rat sarcoma model was created by injecting polymorphic sarcoma 1 (tumor type M1) cells subcutaneously into the inguinal region of the rat. Gold spherical 30-nm particles conjugated to goat antimouse IgG were applied topically for 40 minutes directly to the tumor surface after the skin was incised and reflected away from the tumor site. Next, laser therapy was performed by using laser pulses with the laser beam directed at the center of the tumor. After 24 hours, the rats were euthanized and the tumors excised for evaluation. The results indicated a nonviable (white) area with the diameter close to that of the laser beam. Using clusters of gold nanoparticles clusters, rather than single gold nanoparticles, greatly improved the selectivity of cell killing and enabled the use of lower laser pulse energy, which was less damaging to normal tissues [58].

High-intensity focused ultrasound (HIFU) combined with phase-shift nanoemulsions (PSNEs) is a useful tool for targeting solid tumors. PSNEs are lipid-coated liquid perfluorocarbon droplets that are less than 200 nm. PSNEs can take advantage of the EPR effect of new blood vessels associated with tumor growth. When PSNEs are vaporized into microbubbles in tumors, they enhance HIFU thermal ablation. New Zealand White rabbits were injected with cells of VX2 carcinoma into the thigh muscle. Tumors were grown for 2 weeks. The rabbits were injected with PSNE in the ear vein, and then 2 hours later, HIFU was performed. HIFU heated the PSNEs and ablated the tumor. PSNEs reduced the time and heat intensity needed to cause thermal ablation of solid tumors, thereby improving the feasibility and efficacy of the treatment [59].

Another method for tumor regression is the use of Carboplatin-Fe@C-loaded chitosan nanoparticles, which possess magnetic targeting and heat production properties. They can target liver cancer tissue by using a static magnetic field along with an alternating magnetic field, raise tumor tissue temperature, and facilitate tumor apoptosis [60].

In a study by Li et al. [60], Rat Walker-256 breast carcinoma cells were inoculated intraperitoneally and grown in donor Sprague Dawley infantile rats; these cells were then implanted via laparotomy into the liver of recipient Sprague Dawley rats. If the tumor grew well, the gastroduodenal artery was isolated via a second laparotomy and catheterized, and Carboplatin-Fe@C-loaded chitosan nanoparticles were injected through the gastroduodenal artery into the hepatic artery. Next, the tumor site on the liver was placed in a 5000-Gs static magnetic field for 30 minutes, and then the abdomen was closed. The rat was then placed in the center coils of a high-frequency induction heater. Survival time was the longest for groups treated with Carboplatin-Fe@C-loaded chitosan nanoparticles, magnetic field, and hyperthermia. Carboplatin-Fe@C-loaded chitosan nanoparticles have demonstrated tumor-targeting capability and precise heating of the target site to reach effective therapeutic temperature without injuring the tissue around the target site [60].

Combining magnetization with a functionalized surface, magnetic nanoparticles can selectively attach to the targeted tissue or cell to achieve their therapeutic role. Shakeri-Zadeh et al. [61] injected BALB/c mice subcutaneously into the flank with cells from the mouse colon tumor cell line CT-26, and the tumor was allowed to grow for 14 days. 5-FU is a fluoropyrimidine antimetabolite that is often used to treat colorectal cancer. 5-FU has a short half-life and is known to have adverse effects such as diarrhea, nausea, mouth sores, and low blood counts. Poly lactic-*co*-glycolic acid (PLGA)-coated magnetic nanospheres were used to encapsulate the 5-FU. Magnetic drug targeting was performed by exposing the flanks of the mice to a magnetic field of 0.18T after intravenous injection of the nanocapsule in the mice. The PLGA-coated magnetic nanocapsules with 5-FU inhibited tumor growth when exposed to a magnetic field in mice [61].

The development of nanoparticle-integrin antagonist (IA) as a method of drug delivery is another use of nanotechnology. Therapeutic efficacy of polymerized nanoparticles IA complex that targeted tumor neovasculature was performed by using mice bearing M21-L murine melanoma tumors. The mice were given a single systemic injection of nanoparticle/plasmid complex after the tumors reached 400 mm^3. All of the mice showed >95% tumor reduction, and the majority showed no evidence of the tumor at 6 days after nanoparticle injection [62]. Another model involved murine CT-26 carcinoma cells that were injected intravenously to induce pulmonary metastases or Balb/C mice that were injected intrasplenically to induce hepatic metastases. The nanoparticle gene complexes were given 10 days after injection of the tumor cells. Little to no visible tumor metastases were observed in the liver or lungs of the mice treated with the nanoparticle complex [62].

Treatment efficacy was tested in a study by Li et al. [63] by using subcutaneously injected K1735-M2 (melanoma tumor) in BALB/c mice and CT-26 (colon adenocarcinoma tumor) in C3H mice. The tumors were allowed to reach 150–200 mm^3 before intravenous injection of two nanoparticle complexes to target angiogenesis radiolabeled with Yttrium (^{90}Y). The two complexes were the radiolabeled IA-nanoparticle (IA-NP-^{90}Y) and radiolabeled monoclonal antibody against murine Flk-1 (anti-Flk-1 MAb)-nanoparticle (anti-Flk-1 Mab-NP-^{90}Y) which delayed tumor growth. These radiolabeled nanoparticle complexes were used to target integrin and the VEGFR, which are involved in tumor-induced angiogenesis [63].

Large animal species such as rabbits have been used for nanotechnology projects due to the well-established VX2 tumor model. The VX2 carcinoma (BCRC, Bioresource Collection and Research Center, Taipei, Taiwan) is an anaplastic squamous cell carcinoma derived from a virus-induced papilloma in the wild rabbit but appears as a carcinoma in the domestic species [64]. The VX2 cells may be injected into the organ or muscle of choice to create a tumor at the specific location. The VX2-induced tumor in the leg of a New Zealand White rabbit has been used as a model for nanoparticles incorporating fumagillin, an antimicrobial agent derived from *Aspergillus fumagatus*, to diminish angiogenesis and reduce VX2 tumor growth in rabbits. Fumagillin nanoparticles may be used to suppress the neovasculature and inhibit VX2 adenocarcinoma development using minute drug doses [65].

The science of nanotechnology has been used to create tiny particles or minicells tagged with anticancer antibodies to specifically attach to cancer cells. Minicells are nonliving, anucleate 400-nm particles that are derived from bacteria. Minicell vectors can act as a package to carry chemotherapeutic drugs. Monastrol is a small molecule that arrests mitosis. EGFR was packaged on a minicell with monastral to deliver the drug in a targeted manner. tMDA-MB-468 is a mammary tumor cell line known to overexpress EGFR. Balb/c nu/nu mice were implanted subcutaneously with human breast adenocarcinoma cells between the shoulder blades. After the tumors grew to $\sim 80\ mm^3$, or after 18 days, the mice were treated with $^{EGFR}minicells_{Monastrol}$, administered intravenously. The minicells, derived from *Salmonella typhimurium*, reduced the toxicity associated with chemotherapeutic drugs, allowed reduction of the amount of drug given, and allowed targeting of the tumor [66,67].

In the study by Macdiarmid et al. [66,67] tumor cells were targeted where one arm of the antibody attached to the lipopolysaccharide (LPS) on the minicell surface and the other arm was directed to any tumor cell-surface receptor. Doxorubicin was attached to the minicell ($minicell_{Dox}$). Two dogs with spontaneously occurring advanced (stage IV) T-cell non-Hodgkin lymphoma were injected intravenously with anti-canine-CD3 targeted $minicells_{Dox}$. Rapid tumor regression was noted. The $^{CD3}minicells_{Dox}$ required less doxorubicin per dose than standard multidrug combination chemotherapy without the minicell packaging system. The minicells were well tolerated without adverse effects or deaths. Due to the bacterial origin of the minicells, caution is advised since systemic administration of bacterial products may elicit an inflammatory response. However, no inflammatory processes were noted in this study despite repeated intravenous administration [66].

Specific delivery of chemotherapeutics to tumors reduces the systemic toxicity often attributed to drugs used to treat cancer. Several virus-based nanoparticles can be used to deliver these chemotherapeutics including bacteriophages and plant viruses. One example is the cowpea mosaic virus (CPMV), which is an RNA virus that replicates in back-eyed peas, tobacco, and quinoa. It has been used in nanobiotechnology to deliver targeted drugs to tumors. The structure of the CPMV is similar to that of poliovirus, rhinovirus, and coxsackievirus. To visualize tumor vascularization, human tumor cells embedded in mesh were implanted into chick chorioallantoic membrane and could be tracked over several days, followed by injection of labeled CPMV [68].

To deliver various agents to tumors, the tumor needed to be targeted. M13 bacteriophage has been used to target the vasculature in mice bearing MDA-MB-435–derived (human breast cancer) tumor xenografts using streptavidin-conjugated QDs administered through the tail vein. Other viral particles that have been used in this manner include cowpea chlorotic mottle virus, canine parvovirus, and MS2 bacteriophage [68,69].

CONCLUSIONS

To better understand disease prevention, detection, diagnosis and treatment, living organisms are still required. The use of animal models of cancer and nanoparticles is an ever-changing and dynamic field. As this field continues to evolve, we expect that there will be a trend toward personalized therapies for cancer using engineered nanomaterials. The ability to target specific tissues or trigger drug release at a specific location could reduce the adverse effects of potent chemotherapeutic drugs and allow patients to complete chemotherapeutic regimens that they may not have been able to tolerate previously.

The value of nanotechnology goes beyond its use as a tool for drug delivery. Diagnosis and staging of cancer has the potential to be more accurate using nanotechnology. The use of nanoprobes can be a beneficial prognostic tool. Metastases can be targeted and visualized using nanotechnology. Refinement of the use of nanoparticles and their ability to track tumor cells is continually improving through the utilization of animal models of cancer. Determining if tumor cells have dispersed and traveled to lymph nodes can be more easily detected, thereby allowing physicians to be more knowledgeable about the stage of cancer and determine the best method of treatment.

Nanoparticles can be used as a tool for imaging studies. When seeking to image cancer, contrast agents are often used. An ideal contrast agent would be safe, have minimal adverse effects, have specificity and sensitivity, reach the target quickly, provide high resolution, and be economical.

Treatment of cancer can be more precise using nanoparticles. Tumors are known to have high angiogenic activity and are expected to have more permeable blood vessels, resulting in high intratumoral deposition of nanoscale agents. Furthermore, tumors with leakier blood vessels have the fastest growth rates. Vessel leakiness is strongly associated with the tumor's environment, growth, and rate of angiogenesis. This known trait for these angiogenic blood vessels can be exploited using nanotechnology that has been tested in the animal model.

Nanoparticles have been designed to hone to particular tissues and receptors on tumor cells. This permits targeted therapy that spares healthy tissue and allows lower levels of chemotherapeutic agents to be used. Chemotherapy often has negative side effects that may be irreversible. Animal models of cancer which improve the efficacy of nanoparticle use will provide improved outcomes for cancer patients.

After cancer treatment, nanoprobes can monitor tumor shrinkage. Nanoparticles can also be used to determine if cancer has returned. Detection of cancer may be possible at earlier stages so that better outcomes can be achieved.

Although this chapter was not an exhaustive compilation of all of the possible animal models of cancer, it provides opportunities for the study of cancer, diagnosis, treatment, and post-treatment evaluation using animal models. Animal models run the gamut of mice, rats, rabbits, and dogs and include spontaneous cancer models, genetically engineered cancer models, and xenografts.

REFERENCES

[1] Schroeder A, Heller DA, Winslow MM, Dahlman JE, Pratt GW, Langer R, et al. Treating metastatic cancer with nanotechnology. Nat Rev Cancer 2012;12:39–50.

[2] Caron WP, Lay JC, Fong AM, La-Beck NM, Kumar P, Newman SE, et al. Translational studies of phenotypic probes for the mononuclear phagocyte system and liposomal pharmacology. J Pharmacol Exp Ther 2013;347:599–606.

[3] Milane L, Duan ZF, Amiji M. Pharmacokinetics and biodistribution of lonidamine/paclitaxel loaded, EGFR-targeted nanoparticles in an orthotopic animal model of multi-drug resistant breast cancer. Nanomedicine 2011;7:435–44.

[4] Karathanasis E, Chan L, Karumbaiah L, Mcneeley K, D'orsi CJ, Annapragada AV, et al. Tumor vascular permeability to a nanoprobe correlates to tumor-specific expression levels of angiogenic markers. PLoS ONE 2009;4:e5843.

[5] Benhabbour SR, Luft JC, Kim D, Jain A, Wadhwa S, Parrott MC, et al. In vitro and in vivo assessment of targeting lipid-based nanoparticles to the epidermal growth factor-receptor (EGFR) using a novel Heptameric ZEGFR domain. J Control Release 2012; 158:63–71.

[6] Sarfati G, Dvir T, Elkabets M, Apte RN, Cohen S. Targeting of polymeric nanoparticles to lung metastases by surface-attachment of YIGSR peptide from laminin. Biomaterials 2011;32:152–61.

[7] Tada H, Higuchi H, Wanatabe TM, Ohuchi N. In vivo real-time tracking of single quantum dots conjugated with monoclonal anti-HER2 antibody in tumors of mice. Cancer Res 2007;67:1138–44.

[8] Chien CC, Wang CH, Wang CL, Li ER, Lee KH, Hwu Y, et al. Synchrotron micro-angiography studies of angiogenesis in mice with microemulsions and gold nanoparticles. Anal Bioanal Chem 2010;397:2109–16.

[9] Cousins A, Thompson SK, Wedding AB, Thierry B. Clinical relevance of novel imaging technologies for sentinel lymph node identification and staging. Biotechnol Adv 2014;32:269–79.

[10] Hainfeld JF, Dilmanian FA, Zhong Z, Slatkin DN, Kalef-Ezra JA, Smilowitz HM. Gold nanoparticles enhance the radiation therapy of a murine squamous cell carcinoma. Phys Med Biol 2010;55:3045–59.

[11] Hainfeld JF, Slatkin DN, Focella TM, Smilowitz HM. Gold nanoparticles: a new X-ray contrast agent. Br J Radiol 2006;79:248–53.

[12] Boote E, Fent G, Kattumuri V, Casteel S, Katti K, Chanda N, et al. Gold nanoparticle contrast in a phantom and juvenile swine: models for molecular imaging of human organs using X-ray computed tomography. Acad Radiol 2010;17:410–17.

[13] Liu T, Cousins A, Chien CC, Kempson I, Thompson S, Hwu Y, et al. Immunospecific targeting of CD45 expressing lymphoid cells: towards improved detection agents of the sentinel lymph node. Cancer Lett 2013;328:271–7.

[14] Mcdevitt JL, Mouli SK, Tyler PD, Li W, Nicolai J, Procissi D, et al. MR imaging enables measurement of therapeutic nanoparticle uptake in rat N1-S1 liver tumors after nanoablation. J Vasc Interv Radiol 2014;25:1288–94.

[15] Weissleder R, Elizondo G, Wittenberg J, Lee AS, Josephson L, Brady TJ. Ultrasmall superparamagnetic iron oxide: an intravenous contrast agent for assessing lymph nodes with MR imaging. Radiology 1990;175:494–8.

[16] Yin T, Wang P, Zheng R, Zheng B, Cheng D, Zhang X, et al. Nanobubbles for enhanced ultrasound imaging of tumors. Int J Nanomed 2012;7:895–904.

[17] Hartman KB, Wilson LJ, Rosenblum MG. Detecting and treating cancer with nanotechnology. Mol Diagn Ther 2008;12:1–14.

[18] Cai W, Chen K, Li ZB, Gambhir SS, Chen X. Dual-function probe for PET and near-infrared fluorescence imaging of tumor vasculature. J Nucl Med 2007;48:1862–70.

[19] Cai W, Shin DW, Chen K, Gheysens O, Cao Q, Wang SX, et al. Peptide-labeled near-infrared quantum dots for imaging tumor vasculature in living subjects. Nano Lett 2006;6:669–76.

[20] Yang K, Zhang FJ, Tang H, Zhao C, Cao YA, Lv XQ, et al. In-vivo imaging of oral squamous cell carcinoma by EGFR monoclonal antibody conjugated near-infrared quantum dots in mice. Int J Nanomed 2011;6:1739–45.
[21] Yu X, Chen L, Li K, Li Y, Xiao S, Luo X, et al. Immunofluorescence detection with quantum dot bioconjugates for hepatoma in vivo. J Biomed Opt 2007;12:014008.
[22] Jølck RI, Rydhög JS, Christensen AN, Hansen AE, Bruun LM, Schaarup-Jensen H, et al. Injectable colloidal gold for use in intrafractional 2D image-guided radiation therapy. Adv Healthc Mater 2015;4(6):856–63.
[23] Sakuma S, Yu JY, Quang T, Hiwatari K, Kumagai H, Kao S, et al. Fluorescence-based endoscopic imaging of Thomsen-Friedenreich antigen to improve early detection of colorectal cancer. Int J Cancer 2015;136:1095–103.
[24] Chen MH, Chang CH, Chang YJ, Chen LC, Yu CY, Wu YH, et al. MicroSPECT/CT imaging and pharmacokinetics of 188Re-(DXR)-liposome in human colorectal adenocarcinoma-bearing mice. Anticancer Res 2010;30:65–72.
[25] Cui L, Lin Q, Jin CS, Jiang W, Huang H, Ding L, et al. A PEGylation-free biomimetic porphyrin nanoplatform for personalized cancer theranostics. ACS Nano 2015;9(4):4484–95.
[26] Lovell JF, Jin CS, Huynh E, Jin H, Kim C, Rubinstein JL, et al. Porphysome nanovesicles generated by porphyrin bilayers for use as multimodal biophotonic contrast agents. Nat Mater 2011;10:324–32.
[27] Chanda N, Upendran A, Boote EJ, Zambre A, Axiak S, Selting K, et al. Gold nanoparticle based X-ray contrast agent for tumor imaging in mice and dog: a potential nano-platform for computer tomography theranostics. J Biomed Nanotechnol 2014; 10:383–92.
[28] Nurunnabi M, Khatun Z, Huh KM, Park SY, Lee DY, Cho KJ, et al. In vivo biodistribution and toxicology of carboxylated graphene quantum dots. ACS Nano 2013;7:6858–67.
[29] Brigger I, Dubernet C, Couvreur P. Nanoparticles in cancer therapy and diagnosis. Adv Drug Deliv Rev 2002;54:631–51.
[30] Verdun C, Brasseur F, Vranckx H, Couvreur P, Roland M. Tissue distribution of doxorubicin associated with polyisohexylcyanoacrylate nanoparticles. Cancer Chemother Pharmacol 1990;26:13–18.
[31] Chiannilkulchai N, Ammoury N, Caillou B, Devissaguet JP, Couvreur P. Hepatic tissue distribution of doxorubicin-loaded nanoparticles after i.v. administration in reticulosarcoma M 5076 metastasis-bearing mice. Cancer Chemother Pharmacol 1990; 26:122–6.
[32] Peer D, Margalit R. Tumor-targeted hyaluronan nanoliposomes increase the antitumor activity of liposomal doxorubicin in syngeneic and human xenograft mouse tumor models. Neoplasia 2004;6:343–53.
[33] Kalra AV, Kim J, Klinz SG, Paz N, Cain J, Drummond DC, et al. Preclinical activity of nanoliposomal irinotecan is governed by tumor deposition and intratumor prodrug conversion. Cancer Res 2014;74:7003–13.
[34] Noble CO, Krauze MT, Drummond DC, Forsayeth J, Hayes ME, Beyer J, et al. Pharmacokinetics, tumor accumulation and antitumor activity of nanoliposomal irinotecan following systemic treatment of intracranial tumors. Nanomedicine (Lond) 2014;9:2099–108.

[35] Wang CJ, Zhu GJ, Yu L, Shi BH. Preparation, in vitro, and in vivo antitumor activity of folate receptor-targeted nanoliposomes containing oridonin. Drug Dev Res 2013; 74:43–9.

[36] Goren D, Horowitz AT, Tzemach D, Tarshish M, Zalipsky S, Gabizon A. Nuclear delivery of doxorubicin via folate-targeted liposomes with bypass of multidrug-resistance efflux pump. Clin Cancer Res 2000;6:1949–57.

[37] Stephenson SM, Low PS, Lee RJ. Folate receptor-mediated targeting of liposomal drugs to cancer cells. Methods Enzymol 2004;387:33–50.

[38] Stevens PJ, Sekido M, Lee RJ. A folate receptor-targeted lipid nanoparticle formulation for a lipophilic paclitaxel prodrug. Pharm Res 2004;21:2153–7.

[39] Ryland LK, Doshi UA, Shanmugavelandy SS, Fox TE, Aliaga C, Broeg K, et al. C6-ceramide nanoliposomes target the Warburg effect in chronic lymphocytic leukemia. PLoS ONE 2013;8:e84648.

[40] Tagaram HR, Divittore NA, Barth BM, Kaiser JM, Avella D, Kimchi ET, et al. Nanoliposomal ceramide prevents in vivo growth of hepatocellular carcinoma. Gut 2011;60:695–701.

[41] Adiseshaiah PP, Clogston JD, Mcleland CB, Rodriguez J, Potter TM, Neun BW, et al. Synergistic combination therapy with nanoliposomal C6-ceramide and vinblastine is associated with autophagy dysfunction in hepatocarcinoma and colorectal cancer models. Cancer Lett 2013;337:254–65.

[42] Liu X, Ryland L, Yang J, Liao A, Aliaga C, Watts R, et al. Targeting of survivin by nanoliposomal ceramide induces complete remission in a rat model of NK-LGL leukemia. Blood 2010;116:4192–201.

[43] Jiang Y, Divittore NA, Kaiser JM, Shanmugavelandy SS, Fritz JL, Heakal Y, et al. Combinatorial therapies improve the therapeutic efficacy of nanoliposomal ceramide for pancreatic cancer. Cancer Biol Ther 2011;12:574–85.

[44] Paolino D, Cosco D, Gaspari M, Celano M, Wolfram J, Voce P, et al. Targeting the thyroid gland with thyroid-stimulating hormone (TSH)-nanoliposomes. Biomaterials 2014;35:7101–9.

[45] Tsai CC, Chang CH, Chen LC, Chang YJ, Lan KL, Wu YH, et al. Biodistribution and pharmacokinetics of 188Re-liposomes and their comparative therapeutic efficacy with 5-fluorouracil in C26 colonic peritoneal carcinomatosis mice. Int J Nanomed 2011;6:2607–19.

[46] Drummond DC, Noble CO, Guo Z, Hayes ME, Connolly-Ingram C, Gabriel BS, et al. Development of a highly stable and targetable nanoliposomal formulation of topotecan. J Control Release 2010;141:13–21.

[47] Wang C, Long C, Xie C, Chen X, Zhang L, Chu B, et al. Two novel nanoscale preparations of micelle and thermosensitive hydrogel for docetaxel to treat malignant tumor. J Biomed Nanotechnol 2013;9:357–66.

[48] Berlin JM, Leonard AD, Pham TT, Sano D, Marcano DC, Yan S, et al. Effective drug delivery, in vitro and in vivo, by carbon-based nanovectors noncovalently loaded with unmodified Paclitaxel. ACS Nano 2010;4:4621–36.

[49] Gelperina SE, Khalansky AS, Skidan IN, Smirnova ZS, Bobruskin AI, Severin SE, et al. Toxicological studies of doxorubicin bound to polysorbate 80-coated poly(butyl cyanoacrylate) nanoparticles in healthy rats and rats with intracranial glioblastoma. Toxicol Lett 2002;126:131–41.

[50] Ozpolat B, Sood AK, Lopez-Berestein G. Nanomedicine based approaches for the delivery of siRNA in cancer. J Intern Med 2010;267:44–53.
[51] Matters GL, Harms JF, Mcgovern CO, Jayakumar C, Crepin K, Smith ZP, et al. Growth of human pancreatic cancer is inhibited by down-regulation of gastrin gene expression. Pancreas 2009;38:e151–61.
[52] Shim G, Choi HW, Lee S, Choi J, Yu YH, Park DE, et al. Enhanced intrapulmonary delivery of anticancer siRNA for lung cancer therapy using cationic ethylphosphocholine-based nanolipoplexes. Mol Ther 2013;21:816–24.
[53] Xue W, Dahlman JE, Tammela T, Khan OF, Sood S, Dave A, et al. Small RNA combination therapy for lung cancer. Proc Natl Acad Sci USA 2014;111:E3553–61.
[54] Zeng L, Li J, Li J, Zhang Q, Qian C, Wu W, et al. Effective suppression of the Kirsten rat sarcoma viral oncogene in pancreatic tumor cells via targeted small interfering RNA delivery using nanoparticles. Pancreas 2015;44:250–9.
[55] Rengan AK, Bukhari AB, Pradhan A, Malhotra R, Banerjee R, Srivastava R, et al. In vivo analysis of biodegradable liposome gold nanoparticles as efficient agents for photothermal therapy of cancer. Nano Lett 2015;15:842–8.
[56] Hainfeld JF, O'connor MJ, Lin P, Qian L, Slatkin DN, Smilowitz HM. Infrared-transparent gold nanoparticles converted by tumors to infrared absorbers cure tumors in mice by photothermal therapy. PLoS ONE 2014;9:e88414.
[57] Hainfeld JF, Slatkin DN, Smilowitz HM. The use of gold nanoparticles to enhance radiotherapy in mice. Phys Med Biol 2004;49:N309–15.
[58] Hleb EY, Hafner JH, Myers JN, Hanna EY, Rostro BC, Zhdanok SA, et al. LANTCET: elimination of solid tumor cells with photothermal bubbles generated around clusters of gold nanoparticles. Nanomedicine (Lond) 2008;3:647–67.
[59] Kopechek JA, Park EJ, Zhang YZ, Vykhodtseva NI, Mcdannold NJ, Porter TM. Cavitation-enhanced MR-guided focused ultrasound ablation of rabbit tumors in vivo using phase shift nanoemulsions. Phys Med Biol 2014;59:3465–81.
[60] Li FR, Yan WH, Guo YH, Qi H, Zhou HX. Preparation of carboplatin-Fe@C-loaded chitosan nanoparticles and study on hyperthermia combined with pharmacotherapy for liver cancer. Int J Hyperthermia 2009;25:383–91.
[61] Shakeri-Zadeh A, Shiran MB, Khoee S, Sharifi AM, Ghaznavi H, Khoei S. A new magnetic nanocapsule containing 5-fluorouracil: in vivo drug release, anti-tumor, and pro-apoptotic effects on CT26 cells allograft model. J Biomater Appl 2014;29:548–56.
[62] Guccione S, Li KC, Bednarski MD. Molecular imaging and therapy directed at the neovasculature in pathologies. How imaging can be incorporated into vascular-targeted delivery systems to generate active therapeutic agents. IEEE Eng Med Biol Mag 2004;23:50–6.
[63] Li L, Wartchow CA, Danthi SN, Shen Z, Dechene N, Pease J, et al. A novel antiangiogenesis therapy using an integrin antagonist or anti-Flk-1 antibody coated 90Y-labeled nanoparticles. Int J Radiat Oncol Biol Phys 2004;58:1215–27.
[64] Chen JH, Lin YC, Huang YS, Chen TJ, Lin WY, Han KW. Induction of VX2 carcinoma in rabbit liver: comparison of two inoculation methods. Lab Anim 2004;38:79–84.
[65] Winter PM, Schmieder AH, Caruthers SD, Keene JL, Zhang H, Wickline SA, et al. Minute dosages of alpha(nu)beta3-targeted fumagillin nanoparticles impair Vx-2 tumor angiogenesis and development in rabbits. FASEB J 2008;22:2758–67.

[66] Macdiarmid JA, Madrid-Weiss J, Amaro-Mugridge NB, Phillips L, Brahmbhatt H. Bacterially-derived nanocells for tumor-targeted delivery of chemotherapeutics and cell cycle inhibitors. Cell Cycle 2007;6:2099–105.

[67] Macdiarmid JA, Mugridge NB, Weiss JC, Phillips L, Burn AL, Paulin RP, et al. Bacterially derived 400 nm particles for encapsulation and cancer cell targeting of chemotherapeutics. Cancer Cell 2007;11:431–45.

[68] Manchester M, Singh P. Virus-based nanoparticles (VNPs): platform technologies for diagnostic imaging. Adv Drug Deliv Rev 2006;58:1505–22.

[69] Ellerby HM, Arap W, Ellerby LM, Kain R, Andrusiak R, Rio GD, et al. Anti-cancer activity of targeted pro-apoptotic peptides. Nat Med 1999;5:1032–8.

CHAPTER

Surgical implications of surface texturing in breast implants*

4

D. Srinivasa[1], B. Blumenauer[2] and M.W. Clemens[2]

[1]*University of Michigan, Ann Arbor, MI, United States* [2]*The University of Texas MD Anderson Cancer Center, Houston, TX, United States*

CHAPTER OUTLINE

INTRODUCTION

The advent of nanotechnology in material engineering for reconstructive surgery has bridged prosthesis construct with biologic consequence. With breast augmentation, the placement of a foreign prosthesis elicits a host tissue response that varies with each patient. This tissue response is inflammatory in nature, causing an influx of cytokines that affect the capsule developing around the implant. Depending on the nature of the capsule, circumferential contraction can result in

*No author has any financial interests in relation to the context of this article.

a cosmetic deformity (spherical implant shape) and oftentimes, pain. The ability to apply nanotechnology and materials engineering to modulate the cellular response has opened a plethora of possibilities. Specifically, surface texturing of implants has directly influenced the maturation of implant engineering and implant nanotechnology.

The prosthesis itself has two major components: the outer shell and the material used to fill the implant. For the purposes of this chapter, we will focus on the outer shell and *texturing* that can be applied to the surface of this shell. The industry has progressed through multiple generations of breast implants, starting in 1962 when Cronin and Gerow [1] commercialized the first silicone implant. The capsule of this first implant was thick, and quickly the demand grew for a softer implant. The perceived pliability of the implant, however, was not only the consequence of the capsule thickness, but also the consequence of the capsule tissue character that forms around the implant. The hypertrophic tissue that can form earned the title "capsular contracture." This dreaded consequence has been attributed to multiple causes, most of which are host dependent [2]. A review of the histologic response to implants shows significant promise in altering the texture of the implant to promote fibroblast growth in a targeted manner. The pursuit of the most natural implant has demanded a progressive industry that embraces nanotechnology.

CAPSULAR CONTRACTION

The reported incidence of capsular contraction is variable, secondary to the subjective nature of the evaluation and reporting, ranging from 5% to 75% [2]. Submuscular placement of a textured implant carries the lowest reported rates of capsular contraction. Although lower contracture rates with texturing is supported via multiple randomized control trials and a metaanalysis by Barnsley et al., submuscular placement has been demonstrated with nonrandomized studies [3]. Stevens et al. reproduced these findings with a 5-year follow-up on the Sientra implant with subgroup analysis for texturing and placement [4]. Table 4.1 reviews the various publications used in the Barnsley metaanalysis that addressed implant type and rates of capsular contracture.

The etiology of capsular contracture is debatable. A common theory implicates fibroblast activity in each patient in response to a foreign body. This process is immune modulated [5]. The interface of immunology on a cellular level with implant engineering is where nanotechnology best augments the specialty of Plastic Surgery.

Table 4.1 Literature Review of Implant Types and Capsular Contracture

Study	Year	Outcome Measure	Sample Size	Implant Type	Placement	RR of Contracture at 1 Year	Length of Follow-Up
Asplund	1996	BAC	104	Third-generation silicone, Dow Corning Silastic II HP/Silastic MSI HP	Submuscular	Not reported	12 months
Burkhardt	1994	Baker	90	Saline, Siltex (Mentor)	Subglandular	18	9–32 months
Burkhardt	1995	Baker	104	Saline, Biocell (McGhan)	Subglandular	1.71	5–40 months
Coleman	1991	Baker	50	Third-generation silicone, Siltex (Mentor)	Subglandular	7.58	12 months
Fagrell	2001	Baker	40	Saline, Siltex (Mentor)	Subglandular	4	5 year 11 months–8 year 4 months
Hakelius	1992	BAC	50	Third-generation silicone, Biocell (McGhan)	Subglandular	23	12 months
Tarpila	1997	Baker	42	Saline, Biocell (McGhan)	Subglandular	1.33	12 months

BAC, breast augmentation classification; RR, relative risk; Capsular Contracture was defined as Baker Grade III or IV contracture.

THE CELLULAR RESPONSE TO IMPLANT SURFACE TEXTURING

It is important to understand why texturing the shell of the implant reduces capsular contraction. Later in this chapter we will review in detail the various texturing modalities available today. Texturing disrupts parallel organization of the collagen fibers that constitute the capsule [6]. Capsules of textured implants are thicker, with increased representation of inflammatory markers. However, the collagen is more erratic. Thus it is the force of parallel vectors that contribute to capsular contraction, more so than the thickness or inflammation [7]. The source of this contractile force comes from myofibroblast proliferation and activity [8].

Myofibroblasts are most active during the active phase of wound healing. During this window, they migrate toward the wound and proliferate. In surgical wounds, wound contracture is credited to these cells [9]. However, in the setting of capsular contraction, these cells act on the developing capsule around the prosthesis to create contracture around the implant itself [6]. Hwang et al. evaluated the tensile strength of capsule specimens taken from capsulectomies for capsular contracture. They also measured myofibroblast quantity within the capsule via histologic analysis. They concluded that patients with more clinically evidence capsular contracture as measured by the Baker Score had capsules with increased number of myofibroblasts and collagen deposition [8].

Microscopic analysis of in vivo cellular adhesion to textured versus smooth implants highlights the differences in tissue response to these prostheses. Notably, alpha-actinin and fibroblast adhesion was upregulated in the capsule of a textured implant. Further, cellular adhesion as a whole was significantly lower in the smooth implant capsule [10]. The nature of the foreign body can elicit implant-specific inflammatory reactions. Multiple studies have investigated the role of surface coating. Bacakova et al. demonstrated that varying surface coating molecules elicit type-specific fibroblast activity [12]. Valencia-Lazcano et al. [10] found increased alpha-actinin levels in textured implants coated with collagen I, fibronectin, and hyaluronic acid. Further, FAK expression was upregulated in this subgroup. FAK specifically modulates cellular adhesion in response to cytoskeletal tension. This is closely linked to alpha-actinin, proteins also integral in cellular adhesion. By altering the extracellular matrix proteins coating the implant, we can functionally alter the cellular adhesion properties during the inflammatory response and even more impressively, gene expression. The same molecular coating applied to smooth implants did not see the exaggerated cellular response that was seen in the textured implants. This supports the theory that surface molecular coating is an adjunct to surface texturing, and both are necessary for optimal cellular modulation.

The microscopic detail in which capsules form, and subsequently contract, provides the platform for nanotechnology in designing breast implants. Host cells respond to the placement of the implant, and modifying how host cells respond to the prosthesis can influence the cosmetic outcome. To successfully predict and alter the host response would allow for us, as surgeons, to minimize one of the most frequent and devastating complications of implant-based breast reconstructive surgery.

THE TEXTURED IMPLANT: HISTORY AND DEVELOPMENT

Although recent microscopic details have shed light on how host tissues respond to these prostheses, the movement in modifying breast implants began in the late 1960s. In 1968 Ashley and Pangman patented the first textured breast implant. This implant was coated with polyurethane foam in order to help maintain the implant's position on the chest wall. These implants had silicone gel and an internal scaffold to help maintain shape [13]. The foam capsule is created by a polymerization of polydiethylene glycol adipate and toluene diisocyanate compounds. This mixture was then secured via silicone adhesive to the single lumen implant. Some double lumen, saline, and gel-saline implants were also coated with polyurethane foam. The polyurethane surface allowed for cellular ingrowth and thus, adherence to the forming inflammatory capsule [14].

Tissue growth research by Bradley et al. suggested that polyurethane foam specifically inhibits fibroblasts and delays the formation of the fibrotic capsule [15]. With the increased use of these foam implants, there was an associated decreased incidence in the development of capsular contracture. Even early in the development of implants, there was growing evidence that fibroblast activity is paramount in capsule formation [16].

The polyurethane foam–textured implant became increasingly popular until it was removed from the United States market in 1991 [17], though it remains a popular implant surfacing choice in South America. Rising concerns regarding foam degradation products and the carcinogenic potential in rat models quickly dissuaded their continued use. Specifically, 2,4-toluene-diamene was reported to be the offending carcinogenic agent in an FDA report released in 1991 [14]. Although the polyurethane foam implant was removed from the market, there was a clear desire for an implant that mimicked the same tissue response: capsular growth that would not lead to contracture around the implant.

To fill this void, there was active pursuit by multiple biomedical companies to create their version of a "textured" implant that would provide the same tissue response as the polyurethane foam implant, without the potential carcinogenic component. The two most commonly used textured implants were introduced by McGhan Corporation (Irvine, CA, USA) and the Mentor Corporation (Santa Barbara, CA, USA). McGhan introduced the Biocell line of macrotextured implants (now owned by Allergan) in 1988 and Mentor released the microtextured Siltex implant in 1989. The terms "macrotextured" and "microtextured" reference the diameter of the surface pores. This diameter ranges anywhere from

Table 4.2 Most Commonly Used Textured Implants

Textured Implants	Year Introduced	Technique	Pore Size (μm)
Silastic MSI (Dow Corning)	1989–90	Laser	500–750
MISTI (Bioplasty)	1987–88	Salt loss	20–800
Biocell (Allergan/McGhan/Inamed)	1986–87	Salt loss (sodium chloride)	500–800
Siltex (Mentor)	1989	Negative contact imprint, "Stamping"	60–275

MSI, microstructured implant; MISTI, molecular impact surface textured implant.

150 to 800 μm [18]. Table 4.2 reviews the most commonly used textured implants over the past 20 years, the method used for texturing, and the pore size. These texturing modalities were applied to all implants including both saline and silicone subtypes, and tissue expanders as well.

In response to the polyurethane controversy, multiple companies released a series of nonpolyurethane containing textured implants. Dow Corning (Midland, MI, USA) released the Silastic Micro-Structured Implant (MSI), McGhan released the Biocell implant, Mentor released the Silitex implant, and Bioplasty released the MISTI. Each of these implants can be categorized based on the method of texturing, and the associated benefits and weaknesses. We will discuss this further after reviewing how modern day implants are textured.

Due to safety concerns, the FDA established a moratorium in 1992 on silicone-filled breast implants until substantial scientific evidence was available that satisfactorily outline the safety and efficacy of their use. As a result, several national and multinational studies conducted over the next decade were virtually unanimous in the conclusion that silicone-filled implants did not cause autoimmune diseases. A few exemptions to the use of silicone implants included use in women undergoing breast reconstruction or revision of existing silicone-filled implants. Later approval of premarket applications by Mentor and later Allergan for the use of their silicone implant variants in conjunction with existing evidence attesting to silicone implant safety led to lifting of the FDA moratorium in 2006.

THE SILICONE SHELL

All implants have similar core components. First, there is a template, or *mandrel*, that serves as scaffolding for the shell. This mandrel is then submerged in liquid silicone. The silicone-coated mandrel is placed in a laminar flow cabinet and allowed to set. Once the setting is complete, the shell is now ready to undergo various surface texturing alterations. Crucial in this process is ensuring even dispersion of heat for successful curing. Again, a laminar flow cabinet is used to best evenly distribute heat application. Regardless of the elicited tissue response, symmetric fibroblast growth is necessary for even settling of the implant. The cured implant is then evaluated to ensure that detailed specifications are met prior to medical use. Evaluation consists of microscopic analysis, thickness measurements, and submergence in water to assess for air extravasation. Varying structural qualities of implant shells are unique to the respective manufacturing companies.

SALT LOSS TECHNIQUE

McGhan introduced the salt loss method of surface macrotexturing, aptly named because the silicone is coated with sodium chloride salt, which is then removed or "lost" during a water wash. The shell still begins with a mandrel, which is dipped in silicone. However, before the silicone-coated mandrel is allowed to cure, the implant is covered with granulated salt. The mandrel then cures in the laminar flow cabinet. Once set, the surface is rinsed with water to remove the salt. This leaves the implant with an irregular surface pitted with cuboid shaped pores varying in size and depth (Fig. 4.1). Further details regarding this surface texturing technique remain protected by the manufacturer for proprietary reasons. This technique results in implant surfaces that are irregular and asymmetric since the salt granules are fine and vary in size and shape. The granules themselves are 300–600 μm in size, and thus the resultant texturing results in crevices mirroring these dimensions.

Subsequent studies utilizing SEM (scanning electron microscopy) images of both the implant shell and the capsule that forms reveal crucial features. The surface pores are similar to the honey-combed structure of the polyurethane-foamed surface implants. The ridges have carrying depths that most commonly span 300–600 μm. Given that average fibroblast size is closer to 25 μm, there is significant potential for cellular ingrowth. The porous surface topography precipitates capsular ingrowth that mirrors the detailed ridging on the shell of the implant [19]. Fig. 4.1A and B reveals the cuboid appearance of the pores in salt loss surfaces for the Biocell textured implant. Later, we will discuss the stamping technique. In comparing images, it is easy to appreciate how the stamping technique reveals a more uniform surface texture. Further, some argue that the irregularity of the salt loss technique allows for greater surface area for cellular ingrowth. The degree of collagen ingrowth that creates the mirror image of the surface topography is characteristic of the Biocell brand and polyurethane-coated implants. Interestingly, additional studies have shown that not only is the fibroblast ingrowth more complex, the resultant capsule does not act like synovium to the degree that is seen with a smooth implant [20]. This synovial type epithelium has been implicated in capsules that form to a greater degree with capsules surrounding smooth implants [21]. As evidenced by multiple studies, the nanotopography of textured salt loss implants affects the fibroblast ingrowth pattern, and histology of the subsequent capsule [5].

FIGURE 4.1

(A) SEM Image, 30× of a Biocell textured implant, (B) scanning electron microscopy (sem) image, 43×, Biocell textured implant (bottom) and the associated tissue capsule interface (top).

STAMPING TECHNIQUE

The "Stamping Technique" was introduced by Mentor in 1989, for surface texturing of their Siltex implant shells. Given the success of the initial polyurethane foam implants in the 1960s, there was a desire to pursue similar surface texturing results without placement of the foam itself since it was removed from the market following an FDA warning. Mentor uses a negative contact imprint microtexturing technique for their implant surfaces. These implants again begin with a silicone mandrel. Prior to curing, this silicone is imprinted with polyurethane foam. The irregular honeycomb pattern of the foam creates a nodular surface of varying depths and widths on the implant surface. The resultant texturing is effectively the negative of the polyurethane foam. As such, it is a close mimic of the foam that achieved initial reduction in capsular contraction during the 1960s.

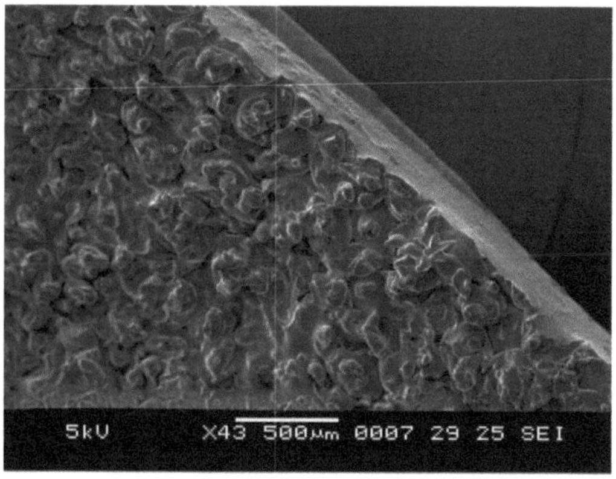

FIGURE 4.2

SEM Image, 43× magnification of a Silitex textured implant formed by negative contact imprinting or "stamping."

Interestingly, microscopic evaluation using SEM shows a linear growth pattern in the fibrous capsule, with 70- to 150-μm diameter surface pores. Fig. 4.2 shows the irregular peaks and valleys seen with magnification of Silitex implants that undergo stamping for surface texturing. One can appreciate the irregular nature of the surface, allowing for a greater amount of surface area for cellular ingrowth. When comparing this with the image of the salt loss technique presented earlier, one can appreciate the size difference between micro- and macrotexturing.

With this method of microtexturing, Barone et al. noted an increase in synovial metaplasia allowing for movement of the implant within the capsule [22]. However, multiple studies including a randomized controlled trial by Malata et al. have shown decreased development of capsular contracture with the Siltex stamping technique compared to smooth surface implants [23]. Overall, texturing of either the macro- or microvariety results in a decreased capsular contracture rate. SEM imaging of the fibrous capsule from stamping technique textured implants do reveal a more organized linear pattern, but whether this translates to a clinically variant rate of capsular contraction when compared to salt loss–curated implants is still debatable.

SURFACE TEXTURING AND CELLULAR INGROWTH

The association between increased capsular contracture with smooth surface implants has been well investigated. At high magnification, smooth surface implants display a paralleled rippling pattern offset by 5-μm groves. Braber et al. correlated parallel surface grooves at 5 μm and less with stronger fibroblast orientation and alignment [24]. It is this very alignment that allows for a concerted contracture, as opposed to more random and disorganized growth. The random growth pattern evades a unidirectional contracture force radially applied to the implant. The alignment of collagen fibers in capsular formation has been associated with the degree of contracture. A metaanalysis by Barnsley et al. demonstrated a reduction in capsular contraction rates in textured breast implants compared to smooth surface implants [3]. Textured surface implants disrupt cellular alignment by allowing for cell ingrowth on the implant surface. The result is capsular conformation to the surface texture of the implant. Bui et al. published a cellular analysis of capsule formation in textured surface implants that demonstrated a lack of myofibroblasts [6]. Myofibroblasts are cells attributed with ability to generate contractile forces in wound healing. These cells act radially on the implant to create the contracture around the implant. The relationship between textured surfaces and lack of myofibroblasts has yet to be clearly identified. It seems that the disorganized collagen and fibroblast growth seen in the capsule of textured implants is paramount.

While microscale surface designs influence cellular response by physical contact, nanoengineering surfaces are thought to alter the manner in which proteins are presented to inflammatory cells [25]. Fibronectin and vitronectin are proteins of the extracellular matrix and are fundamental in cellular adhesion. The size, shape, and depth of nanosurface structures dictate adsorption patterns of these proteins and induce conformational changes in the proteins themselves. Because of these conformational changes, cells such as macrophages and fibroblasts respond differently. This blunts the immune response via downregulation of gene expression of inflammatory proteins such as IL-1β, IL-6, TGF-β1, and EGF as well as collagen 1α2 and collagen 3α1 [21]. Kam et al. described a correlation between high contact angles and greater degree of protein adsorption with nanosurface structures compared to smooth surface controls. Additionally, in this study, test substrates with the most nanoscale surface features demonstrated the slowest growth rate in fibroblasts [21]. There are increasing numbers of examples that reflect how surface nanotechnology can directly influence the cellular response in host tissues. These changes can be traced to the gene expression level. It is important to note that a rare form of non-Hodgkin lymphoma now recognized as breast-implant associated anaplastic large cell lymphoma (BI-ALCL) has been described occurring 8–9 years following implantation [26]. Previous studies have proposed a pathogenic mechanism of chronic T-cell stimulation with local antigenic drive ultimately leading to the development of lymphoma and clinical research into this uncommon disease is ongoing to elucidate the etiology.

UTILITY OF NANOTECHNOLOGY

Given the advancements in nanomanufacturing, and with it the potential for improved medical implant design, there is a natural drive to explore the role of nanotopography in immunomodulation. Implant surface texturing in conjunction with the novel use of surface materials that retard the inflammatory response or inhibit bacterial growth are but a few possibilities. A paper by Nasir et al. postulated nanoscale particles on the surface of an implant that are capable of sensing degradation and rupture of the implant. Further, coating the implant with surface proteins or pharmaceutical agents that can detect and treat breast cancer has also been proposed [27].

A cell's ability to sense and react to its environment is not limited to the surrounding microtopography. Cell filopodia are capable of sensing surface structures down to 10 nm in size [28]. The current literature on microscale engineering suggests that cells are mechanically constrained by the surface topography. Thus the nanotopography of an implant can alter the cellular growth and behavior. This has been well demonstrated by studies on the effect of texturing an implant and rates of capsular contraction. This is just the beginning however. The immune response to a prosthesis can be far more unfortunate in patients with radiated chest walls or poor tissue construct. Biomedical engineering and nanotechnology has opened the door for implants that are host tissue specific.

THREE-DIMENSIONAL SURFACE PRINTING

Biocompatible implant surface topography, often referred to as "biomimetic," has been shown in multiple studies to direct gene expression in cells, specifically breast-derived fibroblasts (BDFs). These cells blunt the foreign body response and subsequent development of a fibrous capsule. Barr et al. published a study in 2010 that investigated the interaction of breast tissue–derived fibroblasts with novel biomimetic surfaces [29]. The study used 32 different microscopic surface patterns using photolithography. Photolithography is a method of transferring 3D shapes to a silicone wafer. A medical grade silicone is then added to the surface of the wafer, cured, then peeled, leaving the desired surface topographies. After the silicone surfaces were cultured with BDFs, immunofluorescent microscopy was used to observe the BDF cytoskeleton forming proteins [23]. The varying patterns of cellular adhesion achieved with different surface topographies highlighted the ability to direct cellular growth patterns by modifying the silicone microenvironment. As manufacturing techniques continue to advance, countless surface patterns can be tailored to enhance biocompatibility between host tissue and synthetic materials.

A recent paper published out of the Manchester Institute of Biotechnology highlights the development of a new biomimetic silicone surface that mimics the acellular dermal matrix (ADM) topography, thus reducing the foreign body response and degree of capsular contraction [30]. The rationale for this new type of surface texturing is to provide the surrounding tissues of the implant with a more natural microenvironment. The result is a cell response that reduces the amount of proinflammatory cytokines released by fibroblasts while promoting cell adhesion and proliferation. Researchers developed a new 3D fabrication process to replicate the unique surface features of ADM in polydimethylsiloxane (PDMS), which is one of the most commonly used silicone-based organic polymers. Available imaging software pieced together images of ADM that were then converted into a bitmap image used during photolithography. The ADM pattern was then transferred to a silicone wafer that acted as a template to produce ADM PDMS stamps using soft lithography [14].

Exposing the ADM PDMS surface varieties to BDFs demonstrated an increase in cell attachment and proliferation on the ADM PDMS surfaces compared to available smooth and textured implant surfaces at varied time intervals [24]. In addition, the study noted an increased BDF survival on the ADM PDMS surface at every measured time interval. The ADM PDMS surfaces promoted a delayed inflammatory cytokine response with downregulation of IL8, TNFa, TGFB1, and HSP60 cytokines at various time intervals compared to conventional textured or smooth surfaced silicone implants [2]. These observed differences were attributed to the unique micro and nanoscale topography of ADM, and a cell's ability to sense and interact with this surface type in a more natural manner.

Nanolithography is a method of producing nanoscale structures by using a mold with 3D surface designs that is pressed against a superheated cast. This is known as the "resist," or effectively a duplicate of the initial mold. An ion etching process can then be used to remove any excess to ensure correct shapes and depths of the duplicate pattern. Molds are created using electron beam lithography to create nanoscale structures. This type of imprint lithography can create structures with high resolution on the scale of 10 nm [31]. Advancing technology will ultimately enable cost effective manufacturing of materials with resolution of smaller than 10 nm that is replicated on a large scale. This imprint method differs from other lithography techniques such as photolithography (which uses light to transfer a specific pattern to a resist followed by chemical treatments to produce the desired 3D patterns) [25].

SURFACE NANOPARTICLES AND INFECTION PREVENTION

The etiology of capsular contracture is multifactorial. Beyond modulating the host immune response to implant surfaces through micro and nanotopographies, one must also prevent infection by reducing the bacterial load after implant insertion. Implant associated infections after breast surgery ranges from 1.1% to 2.5% [32]. With the improvements made in nanoscale manufacturing, nanostructures on implant surfaces can be impregnated with antibacterial materials such as zinc oxide (ZnO) or silver oxide (AgO) to create a surface environment less prone to infection. The use of ZnO on nanostructured surfaces of orthopedic implants decreases the density and number of colony forming units of *Staphylococcus epidermidis* [33,11]. Surface topography manipulation via nanoscale manufacturing and use of antibacterial nanomaterials are just the beginning on the development of the next generation of textured surface breast implants.

CONCLUSION

Implant texturing first made waves in the late 1960s with the advent of polyurethane foam coating. Since then, degradation hazards and research in nanotechnology have launched a resurgence in texturing modalities. Microscopy and tissue studies have demonstrated that the texturing applied can directly affect host tissue response, including gene modification. The ability to affect the cellular response to a foreign prosthesis opens the door for cell-targeted modifications. Given that the capsule that forms around an implant is of paramount consequence in the cosmetic outcome, capsular modification is warranted. Further applications of nanotechnology include surface coating with substances that affect radiation damage, scar formation, and infection. The possibilities are endless, and implant technology is ever changing.

REFERENCES

[1] Cronin TD, Gerow FJ. Augmentation mammaplasty: a new "natural feel" prosthesis. In: Transactions of the Third International Congress of Plastic and Reconstructive Surgery. Amsterdam: Excerpta Medica; 1963. p. 41.

[2] Adams WP. Capsular contracture: what is it? What causes it? How can it be prevented and managed? Clin Plastic Surg 2009;36(1):119–26.

[3] Barnsley G, Sigurdson L, Barnsley S. Textured surface breast implants in the prevention of capsular contracture among breast augmentation patients: a meta-analysis of randomized controlled trials. Plastic Reconstr Surg 2006;117(7):2182–90.

[4] Stevens WG, et al. Risk factor analysis for capsular contracture. Plastic Reconstr Surg 2013;132(5):1115–23.

[5] Kyle DJT, Bayat A. Enhanced contraction of a normal breast-derived fibroblast-populated three-dimensional collagen lattice via contracted capsule fibroblast-derived paracrine factors: functional significance in capsular contracture formation. Plastic Reconstr Surg 2015;135(5):1413–29.

[6] Bui J, Perry T, Ren C, Nofrey B, Teitelbaum S, Epps D. Histological characterization of human breast implant capsules. Aesthet Plast Surg 2015;39(3):306–15.

[7] Abramo AC, et al. How texture-inducing contraction vectors affect the fibrous capsule shrinkage around breasts implants?. Aesthet Plast Surg 2010;34(5):555–60.

[8] Hwang K, et al. Myofibroblasts and capsular tissue tension in breast capsular contracture. Aesthet Plast Surg 2010;34(6):716–21.

[9] Shin D, Minn KW. The effect of myofibroblast on contracture of hypertrophic scar. Plastic Reconstr Surg 2004;113(2):633–40.

[10] Valencia-Lazcano AA, Alonso-Rasgado T, Bayat A. Physico-chemical characteristics of coated silicone textured versus smooth breast implants differentially influence breast-derived fibroblast morphology and behaviour. J Mech Behav Biomed Mater 2014;40:140–55.

[11] Colon G, Ward BC, Webster TJ. Increased osteoblast and decreased *Staphylococcus epidermidis* functions on nanophase ZnO and TiO_2. J Biomed Mater Res 2006;78 (3):595–604.
[12] Bacakova L, et al. Modulation of cell adhesion, proliferation and differentiation on materials designed for body implants. Biotechnol Adv 2011;29(6):739–67.
[13] Peters WJ. Biomaterials in plastic surgery. Oxford: Woodhead Pub; 2012.
[14] Sinclair T, Kerrigan C, Buntic R, Szycher M. Biodegradation of the polyurethane foam covering of breast implants. Plastic Reconstr Surg 1993;92:1003–14.
[15] Bradley S, White K, Mccay J, Brown R, Musgrove D, Wilson S, et al. Immunotoxicity of 180 day exposure to polydimethylsiloxane (silicone) fluid, gel and elastomer and polyurethane disks in female B6C3F1 mice. Drug Chem Toxicol LDCT Drug Chem Toxicol 1994;17(3):221–69.
[16] Institute of Medicine (US) Committee on the Safety of Silicone Breast Implants, Bondurant S, Ernster V, Herdman R, editors. Safety of silicone breast implants. Washington (DC): National Academies Press (US); 1999.
[17] Handel N, Gutierrez J. Long-term safety and efficacy of polyurethane foam-covered breast implants. Aesthet Surg J 2006;26(3):265–74.
[18] Maxwell G, Scheflan M, Spear S, Nava M, Heden P. Benefits and limitations of macrotextured breast implants and consensus recommendations for optimizing their effectiveness. Aesthet Surg J 2014;34:876–81.
[19] Danino A, Basmacioglu P, Saito S, Rocher F, Blanchet-Bardon C, Revol M, et al. Comparison of the capsular response to the biocell RTV and mentor 1600 siltex breast implant surface texturing: a scanning electron microscopic study. Plastic Reconstr Surg 2001;108:2047–52.
[20] Del Rosario AD, Bui HX, Petrocine S, et al. True synovial metaplasia of breast implant capsules: a light and electron microscopic study. Ultrastruct Pathol 1995;19 (2):83–93.
[21] Raso DS, Crymes LW, Metcalf JS. Histological assessment of fifty breast capsules from smooth and textured augmentation and reconstruction mammoplasty prostheses with emphasis on the role of synovial metaplasia. Mod Pathol 1994;7(3):310–16.
[22] Barone F, Perry L, Keller T, Maxwell G. The biomechanical and histopathologic effects of surface texturing with silicone and polyurethane in tissue implantation and expansion. Plastic Reconstr Surg 1992;90:77–86.
[23] Malata C, Feldberg L, Coleman D, Foo I, Sharpe D. Textured or smooth implants for breast augmentation? Three year follow-up of a prospective randomised controlled trial. Br J Plastic Surg 1997;50:99–105.
[24] Braber E, Ruijter J, Smits H, Ginsel L, Recum A, Jansen J. Effect of parallel surface microgrooves and surface energy on cell growth. J Biomed Mater Res 1995;29 (4):511–18.
[25] Kam K, et al. The effect of nanotopography on modulating protein adsorption and the fibrotic response. Tissue Eng A 2014;20(1-2):130–8.
[26] Clemens MW, Miranda RN. Coming of age: breast implant–associated anaplastic large cell lymphoma after 18 years of investigation. Clinics Plast Surg 2015;42(4):605–13.
[27] Nasir A, Brenner S. Think small. Ann Plastic Surg 2012;69:580–7.
[28] Dalby M. Investigating the limits of filopodial sensing: a brief report using SEM to image the interaction between 10 nm high nano-topography and fibroblast filopodia. Cell Biol Int 2004;28:229–36.

[29] Barr S, Hill E, Bayat A. Patterning of novel breast implant surfaces by enhancing silicone biocompatibility, using biomimetic topographies. Eplasty 2010;10:e31.
[30] Kyle D, Oikonomou A, Hill E, Bayat A. Development and functional evaluation of biomimetic silicone surfaces with hierarchical micro/nano-topographical features demonstrates favourable in vitro foreign body response of breast-derived fibroblasts. Biomaterials 2015;52:88–102.
[31] Chou SY, Krauss PR, Renstrom PJ. Nanoimprint lithography. J. Vac. Sci. Technol 1996;14(6):4129–33.
[32] Washer LL, Gutowski K. Breast implant infections. Infect Dis Clin North Am 2012;26(1):111–25.
[33] Young WC, Webster TJ. The role of nanomedicine in growing tissues. Ann Biomed Eng 2009;37(10):2034–47.

CHAPTER

5

Nanoscale engineering of silk fibroin scaffold architecture to repair patient-specific cancer defects

L.W. Dunne, T.S. Iyyanki, C.E. Butler and A.B. Mathur

The University of Texas MD Anderson Cancer Center, Houston, TX, United States

CHAPTER OUTLINE

INTRODUCTION

Silk fibroin (SF) is a naturally derived polymer with properties that are disposed for clinical biomedical applications [1–3]. SF is considered biocompatible, biodegradable, permeable to oxygen and water, possesses high tensile strength with flexibility, and shows a low level of inflammatory response in vivo [4]. SF-based materials have been used in tissue engineering for years and developed to repair critical sized cancer defects musculofascia [3], bone [5,6], cartilage [7], nerve defects, and skin wounds and as blood vessel grafts [6–12].

In order to engineer precision scaffolds for in vivo repair, it is important to characterize the molding polymer solution properties with respect to the resultant scaffolds based on structure–property–function relationship of the polymer, which accounts for the role of intermolecular forces of various magnitudes. SF provides a flexible platform to construct scaffolds with microstructure and mechanical properties that can be engineered via solution-based processing. Efforts have been made to better control and improve the properties of SF scaffolds through techniques such as freezing [13], salt leaching [14], gas foaming [14], dielectrophoresis [15], electrospinning [16,17], direct-write assembly [18], and self-assembly [19].

Through learning from nature, factors involved in the process that silkworms use to produce silk may be applied to help design a successful biomimetic method for making strong materials. *Bombyx mori* SF is derived from the silkworm and is composed of a heavy-chain, light-chain, and a sericin coating. Intermolecular forces align heavy-chain fibroin molecules are in a liquid crystalline state and contribute to the tensile properties of the silk fiber. Several factors, including metallic ion content, shear force, fibroin concentration, and pH, are thought to influence the secondary structure (crystalline) and aggregation of heavy-chain SF molecules (liquid crystalline in solution), which further influences the structural and mechanical properties of the extruded silk fiber [20–27]. As SF polymer solution moves through the secretory pathway in silkworms, it undergoes a series of changes, such as shear force, pH drop from 6.8 to 4.8, and an increase in ionic concentration (e.g., [K +] and [Na +]). These changes lead to a sequence of secondary structural changes in SF that result in the strong mechanical properties of silk fibers.

Our current investigation was designed to configure parameters involved in the natural silkworm fiber extrusion process for SF scaffold fabrication in order to develop scaffold architecture per patient-specific need. The fabrication parameters considered here included SF concentration (0.91–3.33% (w/v)), filter size (filtered by 0.22-, 0.45-, 100-μm filters), pH (4, 5.5, 7.4), and ionic concentration (no phosphate-buffered saline (PBS), 0.1× PBS, 1× PBS). The SF polymer aggregate size in solution was determined by dynamic light scattering (DLS). Scaffold was characterized by scanning electron microscopy (SEM), uniaxial tensile test, and Fourier transform infrared (FTIR) spectroscopy.

MATERIALS AND METHODS

SCAFFOLD PREPARATION

The preparation of pure SF (donated by Dr. S. Hudson, North Carolina State University, Raleigh, NC, USA) has been described in detail by Gobin et al. [4]. Briefly, aqueous SF solution was regenerated using a $Ca(NO_3)_2$–MeOH-H_2O system [28]. SF solutions were prepared at different concentrations ranging from 0.91% (w/v) to 3.33% (w/v). SF solutions were filtered using filters of different sizes (100, 0.45, and 0.22 μm). The pH of the SF solution was also adjusted to different values (pH 4, 5.5, and 7.4). PBS without Ca^{2+} or Mg^{2+} (157 mmol/L Na^+, 4.7 mmol/L K^+) was added to the SF solutions at different final PBS concentrations in the SF solution (no PBS, 0.1 × PBS, and 1 × PBS). SF solutions (5 mL) were placed into 35-mm polystyrene Petri dishes and frozen at −80°C overnight within an ethanol bath container. Frozen samples were then lyophilized. After being crystallized with methanol (15 min) and washed with deionized water (15 min, × 3), samples were cut into strips for analysis. Six samples were prepared for each condition.

SF SOLUTION PARTICLE SIZE MEASUREMENT USING DLS

We characterized the size distribution profile of SF proteins in solution under different conditions using DLS. A particle size analyzer (LB-550, Horiba Scientific, Edison, NJ, USA) was used for size distribution measurements.

3D ARCHITECTURE CHARACTERIZATION USING SEM

The three-dimensional (3D) structure of SF scaffolds was characterized using SEM. Scaffold samples were coated under vacuum using a Balzer MED 010 evaporator (Technotrade International, Manchester, NH, USA) with platinum alloy for a thickness of 25 nm. The samples were examined in a JSM-5910 scanning electron microscope (JEOL USA, Peabody, MA, USA) at an accelerating voltage of 5 kV. Pore size, fibril diameter, and fibril density (# of fibrils per unit area) were measured with Image J (National Institutes of Health, Bethesda, MD, USA).

POROSITY MEASUREMENT

The porosity of porous SF scaffolds was measured by liquid displacement [29]. Briefly, a graduated cylinder containing ethanol was utilized. The initial volume was recorded as V_1. At 5 minutes after the addition of the SF scaffolds to the cylinder, the total volume of ethanol and the ethanol-impregnated scaffold was recorded as V_2. The ethanol-impregnated scaffold was then removed from the cylinder and the residual ethanol volume recorded as V_3. The volume difference $(V_2 - V_1)$ was the volume of the scaffold skeleton. The volume of the ethanol held in the foam was determined as the void volume of the foam $(V_1 - V_3)$.

The total volume of the foam was calculated as $V = (V_2 - V_1) + (V_1 - V_3) = V_2 - V_3$. Thus porosity could be defined as $(V_1 - V_3)/(V_2 - V_3)$.

MECHANICAL PROPERTIES MEASUREMENT USING UNIAXIAL TENSILE TESTING

Mechanical properties of SF scaffolds were measured using the Bose Electro Force 3200 system (Bose Corporation, Minnetonka, MN, USA). Samples were tested with a 50 N load cell (Honeywell Sensotec, Columbus, OH, USA) at a strain rate of 500 μm/s. Ultimate tensile strength (UTS), elastic modulus (E), and strain at failure ($\varepsilon_{failure}$) were calculated from stress–strain data.

CONFORMATION ANALYSIS USING FTIR

FTIR spectroscopy was used to detect the conformation changes (from random coil/helical to β-sheet) that occurred during SF scaffold preparation. All infrared spectra were recorded in the range of 650–4000 cm^{-1} using a Nicolet NEXus 870 FTIR spectrometer (Thermo Scientific, Brookfield, WI, USA). Each spectrum was acquired with a resolution of 4 cm^{-1}. Background measurements were subtracted from the sample readings.

STATISTICAL ANALYSIS

Data are presented as means ± standard deviation (SD). Data sets were analyzed using one-way ANOVA test in SigmaStat; p values less than .05 were considered significant.

RESULTS

EFFECT OF SF CONCENTRATION ON SF PARTICLE SIZE IN SOLUTION

Particle size measurements of SF solutions at different concentrations showed that an increase in SF concentration from 0.91% (w/v) to 3.33% (w/v) led to a significant increase in SF particle size from 178.4 to 547.2 nm ($p < .001$) (Table 5.1A).

EFFECT OF SF CONCENTRATION ON SF SCAFFOLD PROPERTIES

Fig. 5.1 shows representative SEM images of SF scaffolds prepared at different concentrations (0.91% (w/v), 1.74% (w/v), and 3.33% (w/v)). Structural properties (porosity, pore size, fibril size, and fibril density) of SF scaffolds were analyzed and are summarized in Table 5.1B. As the SF concentration increased from 0.91% (w/v) to 3.33% (w/v), the structure of SF scaffolds changed from porous with clustered fibrils to primarily a porous structure with few fibrils. The pore

Table 5.1 Effect of SF Solution Concentration on (A) Particle Size, (B) Scaffold Structure, and (C) Scaffold Mechanical Properties

	0.91% (w/v)	1.31% (w/v)	1.74% (w/v)	3.00% (w/v)	3.33% (w/v)
(A) Dynamic light scattering (DLS) was performed to assess the SF particle size as a function of SF solution concentration					
Mean (nm)	178.4	202.6	243.2	439.2[a,b,c]	547.2[a,b,c,d]
SD (nm)	153.5	156.6	181.8	312.6	392.1
N	100	100	100	100	100
(B) 3D Structure of SF scaffolds was characterized by scanning electron microscopy (SEM) to assess the effect of SF concentration					
Structure description	Porous structure with interconnected sheets and fibrils			Porous structure with interconnected sheets	
Porosity (%) (*N*)	92.7 ± 1.8 (3)	92.3 ± 6.7 (3)	92.4 ± 3.0 (3)	89.1 ± 3.2 (3)	90.0 ± 4.2 (3)
Pore size (μm) (*N*)	231.6 ± 21.4 (10)	212.0 ± 23.7 (10)	174.3 ± 8.7 (10)[e,f]	168.6 ± 19.0 (10)[e,f]	141.1 ± 20.5 (10)[e,f,g,h]
Fibril size (μm) (*N*)	3.39 ± 0.57 (10)	3.32 ± 0.77 (10)	3.59 ± 0.68 (10)	0 ± 0 (10)	0 ± 0 (10)
Fibril density (fibril number/mm^2) (*N*)	234 ± 175 (3)	88 ± 26 (3)	400 ± 253 (3)	0 ± 0 (3)	0 ± 0 (3)
(C) Mechanical properties (ultimate tensile strength, UTS; elastic modulus, *E*; strain at failure) were determined from stress–strain curves obtained for the SF scaffolds via uniaxial tensile test to assess the effect of SF concentration					
UTS (kPa)	195.0 ± 19.0	157.9 ± 27.7	117.4 ± 26.1[i]	169.0 ± 30.6	165.3 ± 51.5
E (kPa)	642.0 ± 41.3	503.5 ± 134.6	318.9 ± 96.4[j]	395.8 ± 73.3	446.1 ± 222.3
Strain at failure	0.94 ± 0.11	0.85 ± 0.09	0.85 ± 0.07	0.99 ± 0.10	0.85 ± 1.25
N	6	6	6	6	6

[a] *$p < .001$ as compared to 0.91% (w/v).*
[b] *$p < .001$ as compared to 1.31% (w/v).*
[c] *$p < .001$ as compared to 1.74% (w/v).*
[d] *$p < .05$ as compared to 3.00 % (w/v).*
[e] *$p < .05$ compared with pore size at 0.91% (w/v).*
[f] *$p < .05$ compared with pore size at 1.31% (w/v).*
[g] *$p < .05$ compared with pore size at 1.74% (w/v).*
[h] *$p < .05$ compared with pore size at 3.00% (w/v).*
[i] *$p < .05$ compared with UTS at 0.91% (w/v).*
[j] *$p < .05$ compared with E at 0.91% (w/v).*

FIGURE 5.1

SEM images of SF scaffolds at different concentrations indicate that SF concentration affected the 3D architecture of SF scaffolds.

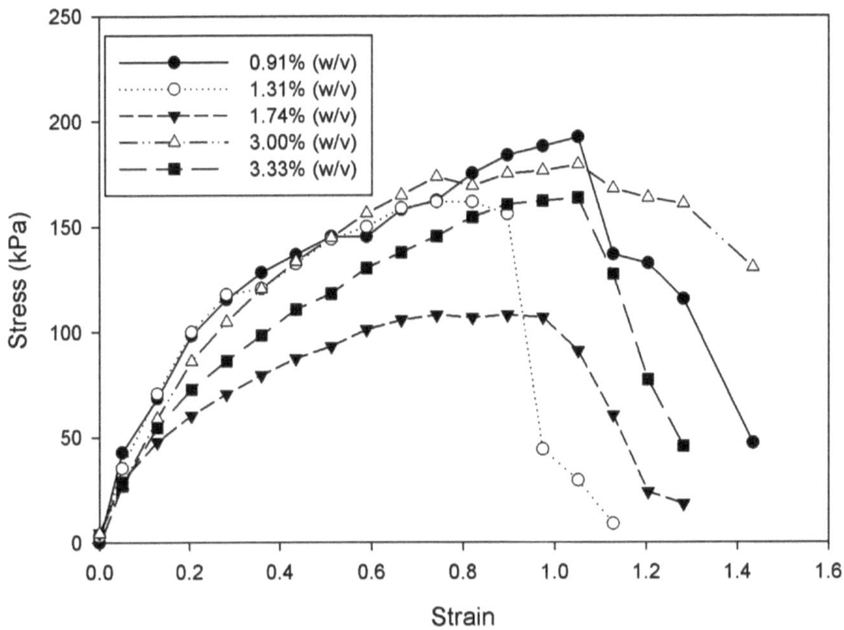

FIGURE 5.2

Representative stress versus strain curves for SF scaffolds with SF concentrations varying from 0.91% (w/v) to 3.33% (w/v).

size of SF scaffolds decreased as the SF concentration increased from 0.91% (w/v) to 3.33% (w/v) (231.6 ± 21.4 vs 141.1 ± 20.5 μm, $p < .05$), while the porosity of SF scaffolds did not change significantly. Fibril structures in SF scaffolds processed from SF solutions at different concentrations (up to 1.74% (w/v)) showed similar sizes and no significant difference in fibril density.

Fig. 5.2 shows the representative stress versus strain curves for SF scaffolds with concentrations varying from 0.91% (w/v) to 3.33% (w/v). Mechanical properties, such as UTS, *E*, and strain at failure, for SF scaffolds made under different conditions are summarized in Table 5.1C. Mechanical properties for SF scaffolds fabricated at a concentration of 1.74% (w/v) showed significantly lower UTS and *E* than SF scaffolds fabricated at other concentrations ($p < .05$).

EFFECT OF THE FILTER SIZE ON SF SCAFFOLD PROPERTIES

Table 5.2 (A–C) shows the effect of the filter size on particle size in SF solution, 3D structure, and mechanical properties of SF scaffolds. Filter of SF solution (1.31% (w/v)) with a filter size 100–0.22 μm led to a significant decrease in the

Table 5.2 Effect of Filter Size on (A) Particle size, (B) Scaffold Structure, and (C) Scaffold Mechanical Properties

	1.31% (w/v) Filtered by 100-μm Filter	**1.31% (w/v) Filtered by 0.45-μm Filter**	**1.31% (w/v) Filtered by 0.22-μm Filter**
(A) SF solution particle size as a function of filter size			
Mean (nm)	202.6	75.8[a]	42.0[a,b]
SD (nm)	156.6	66.9	24.1
N	100	100	100
(B) 3D Structure of SF scaffolds was characterized by SEM to assess the effect of the filter size for 1.31% (w/v) SF solution on the structural properties of SF scaffolds			
Structure description	Porous structure with interconnected sheets and fibrils	Porous structure with interconnected sheets and lots of fibrils	Porous structure with interconnected sheets and fibrils
Porosity (%) (*N*)	92.3 ± 6.7 (3)	94.4 ± 4.1 (3)	91.8 ± 1.3 (3)
Pore size (μm) (*N*)	212.0 ± 23.7 (10)	203.4 ± 29.3 (10)	205.9 ± 17.9 (10)
Fibril size (μm) (*N*)	3.32 ± 0.77 (10)	3.80 ± 0.70 (10)	3.51 ± 0.72 (10)
Fibril density (fibril number/mm^2) (*N*)	88 ± 26 (3)	188 ± 105 (3)	87 ± 44 (3)
(C) Mechanical properties (ultimate tensile strength, UTS; elastic modulus, *E*; strain at failure) were determined from stress–strain curves obtained for the SF scaffolds via uniaxial tensile test to assess the effect of filter size			
UTS (kPa)	157.9 ± 26.7	251.9 ± 67.8[c]	190.4 ± 45.6
E (kPa)	503.5 ± 134.6	885.2 ± 299.3[e]	529.2 ± 93.7[d]
Strain at failure	0.85 ± 0.09	0.71 ± 0.17	1.00 ± 0.17[f]
N	6	6	6

[a] *p < .001 as compared to 100-μm filter.*
[b] *p < .05 as compared to 0.45-μm filter.*
[c] *p < .05 compared with100-μm filter filtered in UTS.*
[d] *p < .05 compared with 0.45-μm filter filtered in E.*
[e] *p < .05 compared with 100-μm filter filtered in E.*
[f] *p < .05 compared with 0.45-μm filter filtered in strain at failure.*

FIGURE 5.3

SEM images of SF scaffolds prepared from SF solutions (1.31% (w/v)) filtered with 100-, 0.45-, or 0.22-μm filters. The 3D structure of SF scaffolds varied with the filter size. Abundant fibril structure appeared in SF scaffolds when the solution was filtered using a 0.45-μm filter.

particle size (reduction from 202.6 to 42.0 nm) (Table 5.2A). SF scaffolds made from solutions filtered using 100- and 0.22-μm filters did not show significant differences in 3D structures with respect to porosity, pore size, fibril size, or fibril density (Table 5.2B). However, SF scaffolds made from solution filtered using the 0.45-μm filter showed many well-defined fibrils and a relatively higher fibril density compared with those sized with 100- and 0.22-μm filters (Fig. 5.3). As shown in Table 5.2C, SF scaffolds made from solution sized using the 0.45-μm filter had a higher UTS and E than scaffolds made from solutions sized using 100- and 0.22-μm filters ($p < .05$).

EFFECT OF pH ON SF SCAFFOLD PROPERTIES

The pH of the SF solution was varied from 4 to 7.4 leading to a increase in SF solution particle size from 107.7 to 289.8 nm significantly ($p < .001$) increased with pH of SF solution (Table 5.3A). SF scaffolds made from pH 5.5 and 7.4 SF solutions appeared to have similar porous structural features, whereas the scaffold produced with pH 4 SF solution had fibrillar morphology (Figure 5.4). Scaffolds produced from pH 4 solution contained many fibrils (Table 5.3B). On the other hand, quantitative analysis showed that the fibril density was higher and fibril size was finer in SF scaffolds at pH 4 than at 5.5 and 7.4. As the pH of the SF solution was increased from 4 to 7.4, the resultant SF scaffolds all had highly porous structures, although the pore size decreased when the pH increased from 5.5 to 7.4 ($p < .05$). Changes in pH of SF solution also affected the mechanical properties of resultant SF scaffolds (Table 5.3C). SF scaffolds made from SF solution at pH 4 presented higher UTS than those made from SF solution at pH 5.5 and 7.4 ($p < .05$).

EFFECT OF PBS CONCENTRATION ON SF SCAFFOLD PROPERTIES

Particle size of SF solutions (1.31% (w/v)) varied with different PBS concentrations (no PBS, 0.1 × PBS, and 1 × PBS) (Table 5.4A). SF solution with no PBS

Table 5.3 Effect of pH on (A) Particle Size, (B) Scaffold Structure, and (C) Scaffold Mechanical Properties

	pH 4	pH 5.5	pH 7.4
(A) SF solution particle size as a function of pH			
Mean (nm)	107.7	202.6[a]	289.8[a,b]
SD (nm)	50.1	156.6	223.7
N	100	100	100
(B) 3D Structure of SF scaffolds was characterized by scanning electron microscopy (SEM) to assess the effect of pH in SF solutions on the structural properties of SF scaffolds			
Structure description	Porous structure with interconnected sheets and lots of fibrils	Porous structure with interconnected sheets and fibrils	Porous structure with interconnected sheets and fibrils
Porosity (%) (*N*)	92.8 ± 3.7 (3)	92.3 ± 6.7 (3)	91.8 ± 2.5 (3)
Pore size (μm) (*N*)	194.3 ± 16.3 (10)	212.0 ± 23.7 (10)	174.3 ± 23.4[c] (10)
Fibril size (μm) (*N*)	2.82 ± 0.71 (10)	3.32 ± 0.77 (10)	5.04 ± 1.07[d,e] (10)
Fibril density (fibril number/mm^2) (*N*)	461 ± 142 (3)	88 ± 26[f] (3)	8 ± 0[f] (3)
(C) Mechanical properties (UTS, ultimate tensile strength; *E*, elastic modulus; strain at failure) were determined from stress–strain curves obtained for the SF scaffolds via uniaxial tensile test to assess the effect of pH in SF solutions			
UTS (kPa)	254.3 ± 39.2	157.9 ± 27.7[g]	185.1 ± 38.3[g]
E (kPa)	577.8 ± 252.0	503.5 ± 134.6	338.0 ± 93.1
Strain at failure	1.00 ± 0.19	0.85 ± 0.09[h]	1.11 ± 0.17
N	6	6	6

[a]*p < .001 as compared to pH 4.*
[b]*p < .001 as compared to pH 5.5.*
[c]*p < .001 compared with pore size at pH 5.5.*
[d]*p < .001 compared with fibril size at pH 4.*
[e]*p < .001 compared with fibril size at pH 5.5.*
[f]*p < .05 compared with fibril density at pH 4.*
[g]*p < .05 compared with before filtering pH 4 in UTS.*
[h]*p < .05 compared with before filtering pH 7.4 in strain at failure.*

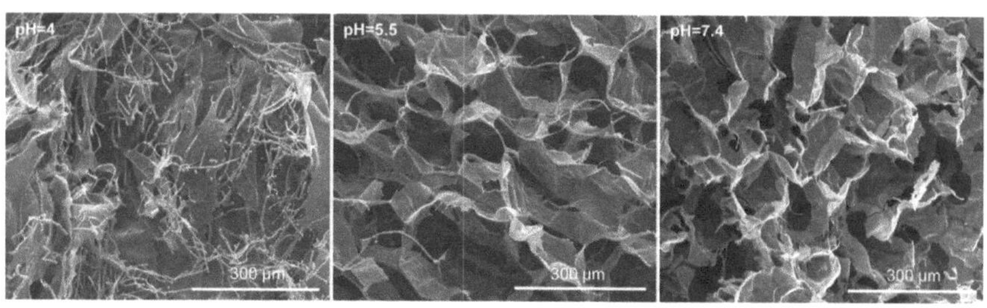

FIGURE 5.4

SEM images of SF scaffolds prepared from SF solutions (1.31% (w/v)) at different pH values. The 3D structure of SF scaffolds varied with pH values in SF solution. Abundant fibril structure appeared in SF scaffolds when the SF solution had a pH value of 4.

Table 5.4 Effect of PBS Concentration on (A) Particle Size, (B) Scaffold Structure, and (C) Scaffold Mechanical Properties

	No PBS	0.1 × PBS	1 × PBS
(A) Dynamic light scattering (DLS) was conducted to assess the SF particle size in 1.31% (w/v) SF solution as a function of PBS concentration			
Mean (nm)	202.6	400.0[a]	252.7
SD (nm)	156.6	536.4	834.9
N	100	100	100
(B) 3D Structure of SF scaffolds was characterized by scanning electron microscopy (SEM) to assess the effect of PBS concentration in SF solution on the 3D structural properties of SF scaffolds			
Structure description	Porous structure with interconnected sheets and fibrils	Porous structure with interconnected sheets and fibrils	Porous structure with interconnected sheets and fibrils
Porosity (%) (*N*)	92.3 ± 6.7 (3)	92.4 ± 3.1 (3)	92.3 ± 4.6 (3)
Pore size (μm) (*N*)	212.0 ± 23.7 (10)	156.2 ± 23.8[b] (10)	146.4 ± 20.5[b] (10)
Fibril size (μm) (*N*)	3.32 ± 0.77 (10)	3.42 ± 0.82 (10)	3.36 ± 0.71 (10)
Fibril density (fibril number/mm^2) (*N*)	88 ± 26 (3)	87 ± 40 (3)	80 ± 17 (3)
(C) Mechanical properties (UTS, ultimate tensile strength; *E*, elastic modulus; strain at failure) were determined from stress–strain curves obtained for the SF scaffolds via uniaxial tensile test to assess the effect of PBS concentration in SF solutions			
UTS (kPa)	157.9.0 ± 27.7	94.7 ± 15.1[c]	190.8 ± 12.9[c,d]
E (kPa)	503.5 ± 134.6	256.1 ± 60.3[e]	469.9 ± 110.0[f]
Strain at failure	0.85 ± 0.09	0.79 ± 0.25	0.96 ± 0.21
N	6	6	6

[a] *$p < .05$ as compared to No PBS.*
[b] *$p < .001$ compared with pore size of the No PBS group.*
[c] *$p < .05$ compared with before filtering No PBS in UTS.*
[d] *$p < .05$ compared with before filtering 0.1 × PBS in UTS.*
[e] *$p < .05$ compared with before filtering No PBS in E.*
[f] *$p < .05$ compared with before filtering 0.1 × PBS in E.*

had a mean particle size of 202.6 nm, whereas SF solutions with final PBS concentrations of 0.1 × PBS and 1 × PBS had mean particle sizes of 400.0 ($p < .05$ as compared to no PBS) and 252.7 nm, respectively. Scaffolds fabricated with SF solutions with or without PBS had primarily porous structures (Fig. 5.5). The presence of PBS in SF solutions did not significantly change the porosity, fibril size, or fibril density in SF scaffolds (Table 5.4B). However, the increase in PBS concentration in SF solution decreased the pore size in SF scaffolds from 212.0 ± 23.7 μm for the no PBS group to 146.4 ± 20.5 μm for the 1 × PBS group ($p < .05$). A change in ionic concentration of SF solution affected the mechanical

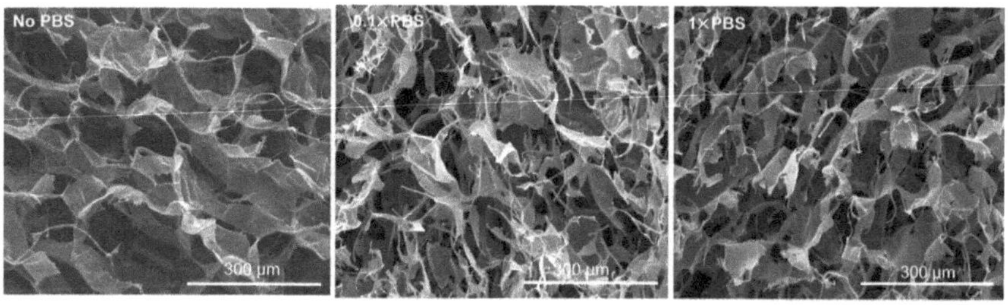

FIGURE 5.5

SEM images of SF scaffolds prepared from SF solutions (1.31% (w/v)) with different PBS concentrations. PBS presence in SF solution and its concentration variance did not affect the 3D structure in SF scaffolds. Thus adjusting the SF solution to a physiologically relevant condition (pH 7.4, isotonic) could be achieved by adding PBS to the SF solution without affecting the 3D structure in the resultant SF scaffolds.

properties of SF scaffolds (Table 5.4C). SF scaffolds made from SF solution with a low PBS concentration (0.1 ×) showed lower UTS and E compared with SF scaffolds made from SF solution without PBS or with a high PBS concentration (1 ×) ($p < .05$).

FTIR PEAK ANALYSIS OF SCAFFOLD STRUCTURE CHANGES DUE TO FILTER SIZE, pH, AND PBS

The secondary structure of SF protein in scaffolds was studied by using FTIR. Fig. 5.6 shows representative FTIR spectra of SF scaffolds made of SF solutions (1.31% (w/v)) at different PBS concentrations. The spectrum revealed the characteristic absorption bands of SF at 1619 cm^{-1} (Amide I), 1513 cm^{-1}, 1523 cm^{-1} (Amide II), and 1232 cm^{-1} (Amide III) that confirm the β-sheet structure. The intensity ratio of β-sheet structure over random structure was calculated to study the relationship between the crystalline and amorphous phases in the SF (Table 5.5B). Table 5.5 (A and B) shows FTIR spectral analysis of SF scaffolds made SF solutions sized with filter sizes, pH, and PBS. Quantitative analysis showed that β-sheet structure was the dominant conformation in all SF scaffolds indicated by Amides I, II, and III. The change in filter size did not affect the ratio of $I_{\beta\text{-sheet structure}}/I_{\text{random structure}}$ in the three amide groups and thus did not affect the primarily crystalline β-sheet secondary structure in the resultant SF scaffolds. The pH of the SF solution did not affect the ratio of $I_{\beta\text{-sheet structure}}/I_{\text{random structure}}$ in the three amide groups of the corresponding SF scaffolds. There was a significant decrease in the ratio of $I_{\beta\text{-sheet structure}}/I_{\text{random structure}}$ in the 0.1 × PBS group compared with the no PBS group (Amide I, $p < .05$).

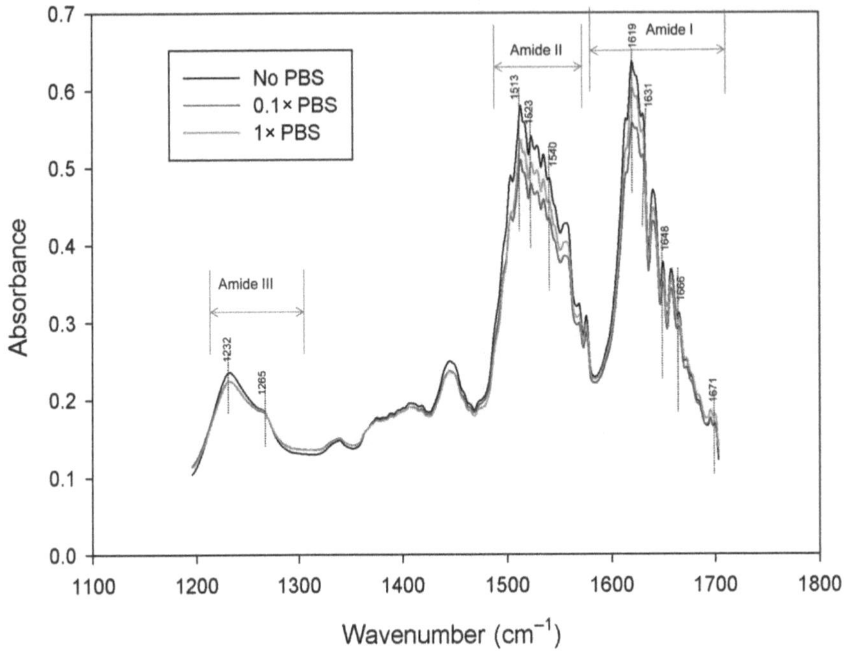

FIGURE 5.6

FTIR spectra of SF scaffolds prepared from SF solutions with different PBS concentrations. Bands for Amide I, Amide II, and Amide III are shown.

DISCUSSION

We conducted this study to understand the correlation between SF solution parameters and the resultant properties of the SF scaffold. The fabrication parameters that were evaluated are SF solution concentration, solution filter size, pH of the solution, and solution ion concentration. The 3D structure and mechanical properties of SF scaffolds were evaluated with respect to these parameters.

The dynamics and aggregation properties of native aqueous SF solutions in the silk gland of *B. mori* silkworms have been studied [20,28]. Several factors, including metallic ion content, shear force, fibroin concentration, and pH, are thought to influence the secondary structure and aggregation of heavy-chain SF molecules [20–27]. The SF solution properties that are affected by these factors also affect the functional properties of the exuded SF fiber from the silk gland. These factors have also been explored in regenerated SF solution for biomedical applications [13,18,30–35]. The same factors alter the solution properties of dissolved SF extracted from the silk fiber and affect the structure and mechanical properties of the resulting 3D SF scaffolds as explored in this study.

Table 5.5 FTIR Spectra Analysis for SF Scaffolds

Designation	Approximate Frequency (cm^{-1})	Wavenumber of Observed Peaks (cm^{-1})	Description
(A) Characteristic infrared bands and wavenumbers of SF scaffolds			
Amide I	1600–1690	1619, 1625, 1631	β-Sheet structure
		1648, 1666, 1671	Random structure
		1672, 1677	β Turn
		1650, 1656	α Structure
Amide II	1480–1575	1513, 1517, 1523, 1529	β-Sheet structure
		1540, 1546	Random structure
Amide III	1229–1301	1232	β-Sheet structure
		1265	Random structure

	Filter Size			pH			PBS		
(B) Summarization of FTIR spectra analysis for SF scaffolds as a function of filter size, pH, and ionic concentration									
$I_{\beta\text{-sheet structure}}/I_{\text{random structure}}$	0.22 μm	0.45 μm	100 μm	pH 4	pH 5.5	pH 7.4	No PBS	0.1 × PBS	1 × PBS
Amide I ($I_{1619\ cm^{-1}}/I_{1671\ cm^{-1}}$)	2.4 ± 0.1	2.5 ± 0.1	2.6 ± 0.1	2.4 ± 0.1	2.6 ± 0.1	2.4 ± 0.1	2.6 ± 0.1	2.3 ± 0.1[a]	2.4 ± 0.1
Amide II ($I_{1523\ cm^{-1}}/I_{1540\ cm^{-1}}$)	1.1 ± 0.0	1.1 ± 0.0	1.1 ± 0.0	1.1 ± 0.0	1.1 ± 0.0	1.1 ± 0.0	1.1 ± 0.0	1.1 ± 0.0	1.1 ± 0.0
Amide III ($I_{1232\ cm^{-1}}/I_{1265\ cm^{-1}}$)	1.3 ± 0.0	1.3 ± 0.0	1.3 ± 0.0	1.3 ± 0.0	1.3 ± 0.0	1.3 ± 0.0	1.3 ± 0.0	1.3 ± 0.0	1.3 ± 0.0
N	3	3	3	3	3	3	3	3	3

[a] *$p < .05$ compared with no PBS in Amide I group.*

In vitro studies have shown that SF concentration affects aggregation of SF polymer chains in solution and SF gel formation [28,31]. In this study, the increase in SF concentration resulted in an increase in the measured particle size of SF in solution. This apparent increase in particle size could be a result of aggregation of SF polymer chains in solution and detected as larger particles due to higher SF content [28].

In this study SF concentration affected 3D architecture of SF scaffolds. As the SF concentration increased (from 0.91% (w/v) to 3.33% (w/v)), the pore size increased and the SF scaffold structure changed from fibril structures to sheet structures. Other studies have shown that an increase in SF concentration from 4% (w/v) to 20% (w/v) lessens the SF gelation time, which affects the resulting material structure (smaller pore size, increase in compression modulus) due to aggregation of closely associated polymer chains at higher concentration in solution, a common phenomenon observed for SF in solution [28].

Under gland mimicking solution conditions, a smaller ($p < .001$) solution particle size is observed at pH 4 that lead to smaller fibril size, a higher fibril density, and higher UTS of the SF scaffolds; however the dominant conformation in SF scaffolds at all three pHs is β-sheets. In the presence of fibrillar scaffold structure, the polymer conformation is β-sheets. In the natural silk-spinning process, there is a reduction in pH (pH 4.9) in the gland lumen as fibroin solution passes from the posterior division through the secretory pathway to the anterior division resulting in liquid crystalline ordering of the SF polymer chains and β-sheet fiber structure [23,25]. Studies have shown that pH changes in SF solution would lead to the transformation of spherical micelles at pH 6.8 to nanofibrils at pH 4.8 [35]. It is possible that at low pH, strong interchain aggregation [33,34], chain entanglement, higher order assembly of polymer chains [28] can accelerate or lead to gelation of SF solution under extreme conditions [31] or as in this study it lead to finer changes in scaffold architecture.

As PBS (with major Na^+ and K^+) concentrations varied, the 3D structure of SF scaffolds did not change significantly except for a decrease in pore size with increased ionic concentration and changes in mechanical properties. An increase in the concentrations of Na^+ and K^+ by adjusting PBS concentration from no PBS to 0.1 × lead to a decrease in UTS and elastic modulus, probably due to the breakdown of inter- and intrapolymeric forces that are perturbed by the initial insertion of ions in to the solution. Further increase (0.1 × to 1 × PBS) in ionic concentration past a threshold increased the UTS and elastic modulus. The effect of K^+ ions on SF conformation in the silkworms has been reported. High K^+ ion content was found in the distal end of silk-secreting ducts in silkworms, which contributed to changes in secondary structure from random coil to β-sheet [28,36]. Zhou et al. [24] reported that increased Na^+ concentration also induced the conversion from random coil to β-sheet. SF scaffolds made from 1 × PBS groups have higher UTS and elastic modulus than the 0.1 × PBS group, which is attributed to the significantly higher β-sheet content observed in 0.1 × PBS scaffolds [30].

Protein conformational changes can be induced by various factors, such as shear force, pH, metallic ions, and others. Air-dried SF membranes made from regenerated

SF aqueous solution using a $Ca(NO_3)_2$-MeOH-H_2O system or LiBr-EtOH-H_2O system at room temperature showed mainly random coil conformation [37]. Characteristic peaks at 1657, 1651 cm^{-1} (Amide I), 1535 cm^{-1} (Amide II), and 1238 cm^{-1} (Amide III) were observed in these untreated membranes [37]. Ethanol treatment of SF membranes made from a LiBr-EtOH-H_2O system showed an increasing band at 1618 cm^{-1}, indicating β-sheet formation coincident with a decreasing band at 1668 cm^{-1}, indicating loss of random coil structure [27]. SF scaffolds fabricated in this study ($Ca(NO_3)_2$-MeOH-H_2O, freeze-drying, and methanol crystalization) showed strong signals of characteristic peaks at 1619 cm^{-1} (Amide I), 1513 cm^{-1}, 1523 cm^{-1} (Amide II), and 1232 cm^{-1} (Amide III) that indicated β-sheet structure formation. Quantitative analysis indicated that β-sheet structure is the dominant conformation in SF scaffolds in the Amide I, II, and III groups for all the studied SF scaffolds. Methanol treatment is effective in inducing β-sheet formation in regenerated SF membranes [32]. It is possible that after crystallization by methanol, effects from the filter size, pH adjustment, and PBS presence on characteristic peaks in the final SF scaffolds were not detectable in this study. However, a slight difference in the $I_{\beta\text{-sheet structure}}/I_{\text{random structure}}$ ratio in Amide I was observed in SF scaffolds made with SF solutions of different PBS concentrations. It is possible that the ion effect on conformation change is more dominant in SF scaffolds.

The SF scaffold fibril size was affected by the particle size of the solution that was varied by changing the concentration, filter size, pH, and ionic concentration. A plot of fibril size as a function of particle size for all solution treatment conditions shows the spread of particle sizes from 40 to 550 nm (Fig. 5.7). The smaller filter sizes and low pH resulted in smaller particle sizes. Larger particle sizes

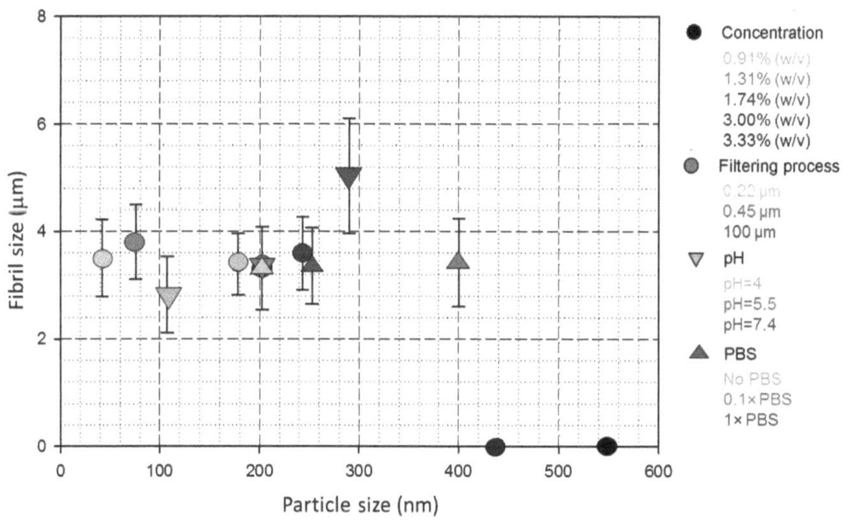

FIGURE 5.7

Relationship between SF solution properties and the resulting scaffold architecture.

were observed for higher SF concentration solution. This data was compiled to show one aspect of the scaffold microstructure that is affected by polymer solutions properties and thus provides a measure and means for scaffold design per clinical applications.

CONCLUSION

The 3D structure and mechanical properties of SF scaffolds are tunable by controlling the SF concentration, filter size, ion concentration, and pH of SF solutions. Structural properties of SF scaffold such as fibril versus sheet structure and fibril density affects the mechanical properties of SF scaffold. The primary secondary structure β-sheet conformation contributes to higher mechanical properties in the scaffolds. On the basis of this study, factors involved in the natural silk-spinning process could be utilized to provide a biomimetic strategy to improve mechanical and architectural properties of the SF scaffold. Strong SF scaffolds with defined porous and microfibrous structures can be made through adjusting SF concentration, pH, and ion concentration and can be tuned per clinical need and application.

ACKNOWLEDGMENTS

This study was supported by a National Institutes of Health grant (NIHR01 AG034658). We thank Dr. Hudson (North Carolina State University) for the donation of raw silk. We also thank the High Resolution Electron Microscopy Facility (HREMF; Cancer Center Core Grant CA16672) for SEM imaging.

REFERENCES

[1] Altman GH, Diaz F, Jakuba C, Calabro T, Horan RL, Chen JS, et al. Silk-based biomaterials. Biomaterials 2003;24:401–16.

[2] Panilaitis B, Altman GH, Chen J, Jin HJ, Karageorgiou V, Kaplan DL. Macrophage responses to silk. Biomaterials 2003;24:3079–85.

[3] Clemens MW, Downey S, Agullo F, Lehfeldt MR, Kind GM, Palladino H, et al. Clinical application of a silk fibroin protein biologic scaffold for abdominal wall fascial reinforcement. Plastic Reconstr Surg Global Open 2014;2:e246.

[4] Gobin AS, Froude VE, Mathur AB. Structural and mechanical characteristics of silk fibroin and chitosan blend scaffolds for tissue regeneration. J Biomed Mater Res A 2005;74:465–73.

[5] Rios CN, Skoracki RJ, Mathur AB. GNAS1 and PHD2 short-interfering RNA support bone regeneration in vitro and in an in vivo sheep model. Clin Orthop Relat Res 2012;470:2541–53.

[6] Rios CN, Skoracki RJ, Miller MJ, Satterfield WC, Mathur AB. In vivo bone formation in silk fibroin and chitosan blend scaffolds via ectopically grafted periosteum as a cell source: a pilot study. Tissue Eng A 2009;15:2717–25.

[7] Zang M, Zhang Q, Davis G, Huang G, Jaffari M, Rios CN, et al. Perichondrium directed cartilage formation in silk fibroin and chitosan blend scaffolds for tracheal transplantation. Acta Biomater 2011;7:3422–31.

[8] Altman AM, Yan Y, Matthias N, Bai X, Rios C, Mathur AB, et al. IFATS collection: human adipose-derived stem cells seeded on a silk fibroin-chitosan scaffold enhance wound repair in a murine soft tissue injury model. Stem Cells 2009;27:250–8.

[9] Gobin AS, Butler CE, Mathur AB. Repair and regeneration of the abdominal wall musculofascial defect using silk fibroin-chitosan blend. Tissue Eng 2006;12: 3383–94.

[10] Lovett M, Cannizzaro C, Daheron L, Messmer B, Vunjak-Novakovic G, Kaplan DL. Silk fibroin microtubes for blood vessel engineering. Biomaterials 2007;28:5271–9.

[11] Uebersax L, Mattotti M, Papaloizos M, Merkle HP, Gander B, Meinel L. Silk fibroin matrices for the controlled release of nerve growth factor (NGF). Biomaterials 2007;28:4449–60.

[12] Yang Y, Ding F, Wu H, Hu W, Liu W, Liu H, et al. Development and evaluation of silk fibroin-based nerve grafts used for peripheral nerve regeneration. Biomaterials 2007;28:5526–35.

[13] Tamada Y. New process to form a silk fibroin porous 3-D structure. Biomacromolecules 2005;6:3100–6.

[14] Nazarov R, Jin HJ, Kaplan DL. Porous 3-D scaffolds from regenerated silk fibroin. Biomacromolecules 2004;5:718–26.

[15] Gupta V, Davis G, Gordon A, Altman AM, Reece GP, Gascoyne PR, et al. Endothelial and stem cell interactions on dielectrophoretically aligned fibrous silk fibroin-chitosan scaffolds. J Biomed Mater Res A 2010;94:515–23.

[16] Min BM, Lee G, Kim SH, Nam YS, Lee TS, Park WH. Electrospinning of silk fibroin nanofibers and its effect on the adhesion and spreading of normal human keratinocytes and fibroblasts in vitro. Biomaterials 2004;25:1289–97.

[17] Zhou SB, Peng HS, Yu XJ, Zheng XT, Cui WG, Zhang ZR, et al. Preparation and characterization of a novel electrospun spider silk fibroin/poly(D,L-lactide) composite fiber. J Phys Chem B 2008;112:11209–16.

[18] Ghosh S, Parker ST, Wang XY, Kaplan DL, Lewis JA. Direct-write assembly of microperiodic silk fibroin scaffolds for tissue engineering applications. Adv Funct Mater 2008;18:1883–9.

[19] Lu Q, Wang X, Lu S, Li M, Kaplan DL, Zhu H. Nanofibrous architecture of silk fibroin scaffolds prepared with a mild self-assembly process. Biomaterials 2011;32:1059–67.

[20] Hossain KS, Ochi A, Ooyama E, Magoshi J, Nemoto N. Dynamic light scattering of native silk fibroin solution extracted from different parts of the middle division of the silk gland of the *Bombyx mori* silkworm. Biomacromolecules 2003;4:350–9.

[21] Ochi A, Hossain KS, Magoshi J, Nemoto N. Rheology and dynamic light scattering of silk fibroin solution extracted from the middle division of *Bombyx mori* silkworm. Biomacromolecules 2002;3:1187–96.

[22] Shao JZ, Zheng JH, Liu JQ, Carr CM. Fourier transform Raman and Fourier transform infrared spectroscopy studies of silk fibroin. J Appl Polym Sci 2005;96:1999–2004.

[23] Terry AE, Knight DP, Porter D, Vollrath F. pH induced changes in the rheology of silk fibroin solution from the middle division of *Bombyx mori* silkworm. Biomacromolecules 2004;5:768–72.
[24] Zhou L, Chen X, Shao ZZ, Huang YF, Knight DP. Effect of metallic ions on silk formation in the mulberry silkworm, *Bombyx mori*. J Phys Chem B 2005;109:16937–45.
[25] Zhou P, Xie X, Deng F, Ping Z, Xun X, Feng D. Effects of pH and calcium ions on the conformational transitions in silk fibroin using 2D Raman correlation spectroscopy and C-13 solid-state NMR. Biochemistry 2004;43:11302–11.
[26] Chen X, Shao ZZ, Knight DP, Vollrath F. Conformation transition kinetics of *Bombyx mori* silk protein. Proteins 2007;68:223–31.
[27] Chen X, Shao ZZ, Marinkovic NS, Miller LM, Zhou P, Chance MR. Conformation transition kinetics of regenerated *Bombyx mori* silk fibroin membrane monitored by time-resolved FTIR spectroscopy. Biophys Chem 2001;89:25–34.
[28] Mathur AB, Tonelli A, Rathke T, Hudson S. The dissolution and characterization of *Bombyx mori* silk fibroin in calcium nitrate methanol solution and the regeneration of films. Biopolymers 1997;42:61–74.
[29] Zhang R, Ma PX. Poly(alpha-hydroxyl acids)/hydroxyapatite porous composites for bone-tissue engineering. I. Preparation and morphology. J Biomed Mater Res 1999;44:446–55.
[30] Kim UJ, Park J, Kim HJ, Wada M, Kaplan DL. Three-dimensional aqueous-derived biomaterial scaffolds from silk fibroin. Biomaterials 2005;26:2775–85.
[31] Kim UJ, Park JY, Li CM, Jin HJ, Valluzzi R, Kaplan DL. Structure and properties of silk hydrogels. Biomacromolecules 2004;5:786–92.
[32] Lawrence BD, Wharram S, Kluge JA, Leisk GG, Omenetto FG, Rosenblatt MI, et al. Effect of hydration on silk film material properties. Macromol Biosci 2010;10:393–403.
[33] Matsumoto A, Chen J, Collette AL, Kim UJ, Altman GH, Cebe P, et al. Mechanisms of silk fibroin sol-gel transitions. J Phys Chem B 2006;110:21630–8.
[34] Cao ZB, Chen X, Yao JR, Huang L, Shao ZZ. The preparation of regenerated silk fibroin microspheres. Soft Matter 2007;3:910–15.
[35] Chen P, Kim HS, Park CY, Kim HS, Chin IJ, Jin HJ. pH-triggered transition of silk fibroin from spherical micelles to nanofibrils in water. Macromol Res 2008;16:539–43.
[36] Ruan QX, Zhou P, Hu BW, Ji D. An investigation into the effect of potassium ions on the folding of silk fibroin studied by generalized two-dimensional NMR-NMR correlation and Raman spectroscopy. FEBS J 2008;275:219–32.
[37] Chen X, Knight DP, Shao ZZ, Vollrath F. Regenerated *Bombyx silk* solutions studied with rheometry and FTIR. Polymer 2001;42:9969–74.

CHAPTER

Magnetic nanoparticles and cancer

6

S. Bucak and C.L. Altan

Yeditepe University, Istanbul, Turkey

CHAPTER OUTLINE

INTRODUCTION

Each year, more than 14 million people are diagnosed with and more than 8 million lose their lives because of different types of cancer. About 24.6 million are still alive among those who have been diagnosed in the past 5 years. In the developing countries, the risk of cancer is predicted to grow until 2020 (~15 million patients), whereas in developed countries this number is expected to remain stable [1]. On the other hand, recent scientific studies give rise to the emergence of novel treatment methods that lead to an increasing amount of patients who had successful treatments in consequence of early diagnosis. However, cancer is still the number one cause of death in the world. Currently, surgery, chemotherapy, immunotherapy, and radiation therapy are the most used treatment methods for cancer but none still offers a true total recovery. In many cases, surgical eradication or radiation treatments are not feasible. On the other hand, many chemotherapeutic drugs are cytotoxic and distributed within the whole body causing toxicity along with serious side effects so that it becomes impossible for the patient to use them [2–4]. These cytotoxic drugs also attack normal and healthy cells in addition to the primary target during the treatment process.

Within the last two decades, magnetic nanoparticles (MNPs) have gained a lot of attention in biomedical applications such as magnetic resonance imaging (MRI) [5–7], tissue repair [8,9], immunoassay [10,11], detoxification of biological fluids [12], magnetic hyperthermia [13–17], drug delivery [18–24], and cell and protein separation [25,26] due to their biocompatibility, facile surface

Nanotechnology in Cancer.

modification, and unique magnetic properties. One of the most important features of MNPs is the possibility of manipulation in the presence of a magnetic field gradient. The other advantages of MNPs in biomedical applications are that they can be used for visualization as in MRI, guided or targeted to a specific location by the application of an external magnetic field followed by induced drug delivery or hyperthermia as a result of heating in an alternating magnetic field causing depredation of cancer cells (Fig. 6.1).

MNPs, although may contain different elements such as cobalt and nickel, are generally composed of iron oxides. In biomedical applications, magnetite (Fe_3O_4) is the most preferred type of iron oxide due to its excellent magnetic properties. It consists of both iron (II) and iron (III) ions and it is the most magnetic, naturally occurring mineral. There are various synthesis methods for the preparation of magnetite nanoparticles which are capable of controlling the particle size ranging from a few nanometers up to tens of nanometers [5,8]. The effective use of magnetite nanoparticles in biomedical applications depend on the size, morphology, and most importantly the magnetic properties of the particles. As the formation mechanism affect these properties, it is necessary to decide the suitable synthesis method for the preparation of magnetite nanoparticles that can be used in specific biomedical applications in order to obtain the most feasible magnetization values along with narrow size distribution thus have uniform physical and chemical properties. Among a variety of synthesis methods, chemical coprecipitation is the most used wet chemical route due to its simplicity and facile reaction kinetics [27]. The essential motive for the use of this method is the ability to synthesize relatively smaller MNPs (5–20 nm) which have superparamagnetic properties.

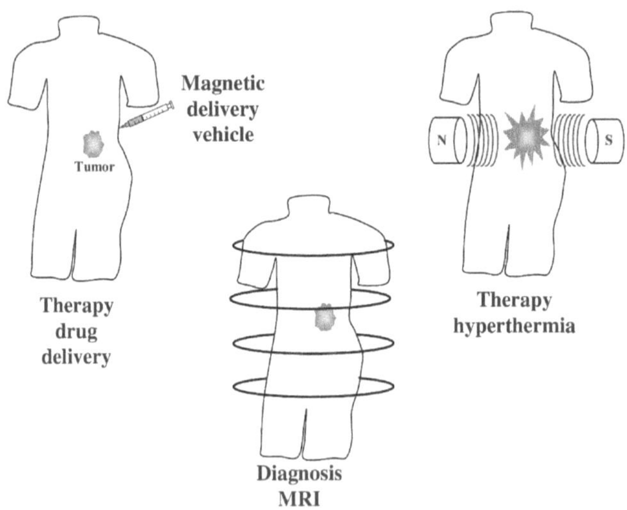

FIGURE 6.1

Different uses of magnetic particles in drug delivery, hyperthermia, and MRI.

As a result of this property, the organization of particles can be manipulated in nanofluids and particles align themselves according to the applied external magnetic field direction while in the absence they do not have any remaining magnetization. Coprecipitation method can also be performed in the presence of suitable surface active agents and the size, colloidal stability and magnetic properties of the particles can easily be modified. These superparamagnetic nanoparticles have wide applications from magnetic drug targeting (MDT) to contrast enhancement in MRI. Another aqueous synthesis method for the preparation of different magnetic iron oxide nanoparticles is partial oxidation by which larger ($\sim$50 nm) single-domain particles that have ferrimagnetic properties can be synthesized [28–33]. As these particles exhibit permanent magnetic properties, it is generally challenging to obtain colloidal stabilization in dispersions. However, due to their enhanced magnetic properties, these particles are suitable candidates for many biomedical applications such as magnetic hyperthermia and cell separations.

As the size of MNPs are smaller or similar to those of a cell, a protein, etc., they can bind to drugs, proteins, enzymes, antibodies, or nucleotides and target a specific region within the body by using an external magnetic field [34]. Furthermore, MNPs can also be functionalized with biological molecules to enhance interaction or binding to an organ, a tissue, or a tumor which provides a controllable tagging [35–37].

As summarized earlier, with excellent magnetic properties which allow their manipulation upon application of an external magnetic field and with tunable sizes that determine their type of magnetism (para-, super-, or ferri-), high surface areas, and functionable surfaces, MNPs lend themselves for various biomedical applications. One very important aspect that should be carefully examined is the biocompatibility of these particles. Most studies in the literature investigate the in vitro toxicity of these particles; however, their applicability depends on their in vivo toxicity [38]. In vitro cytotoxicity test are performed on different cell lines where their viability in the presence of different doses of MNPs are investigated over a short period of time [39,40]. In vivo toxicity test involve the exploration of the distribution of these particles in different tissues and organs [38,41]. An excellent review on the toxicity of these particles is recently published where work in this field is very well organized [42]. In the biomedical field MNPs of a great variety of sizes (few nanometers to micrometers) are employed. These particles may be bare or coated with a variety of coatings such as citric acid, oleic acid, lipids, dextrans, other sugar-based materials, polymers such as PEG, PAA, and PEI or biopolymers such as chitosan and gum arabic or they may be embedded in biodegradable polymers such as PLLA and PLGA, or maybe formulated as hydrogels or microcapsules [5,8,27]. Depending on the application, different doses and exposure times are required. In addition the type of cells where the target is greatly varies. Considering all applications and types of MNPs used, it is impossible to come to a general conclusion about the safety of these particles. In 1996, FDA approved the use of SPIONs (superparamagnetic iron oxide nanoparticles) as MRI agents. However, based on published results in the literature, it is evident that cytotoxicity depends on:

- The cell type exposed to particles where some cell lines are found to be less tolerant [40].
- The type of coating, despite various reports there is somewhat a common understanding that bare particles are least tolerated [43,44].
- The dose, where particles of whichever type is found to be toxic above a threshold concentration, that concentration being dependent of the particle itself and the type of cell line used [38,44].
- The size, which dictates where these particles will be accumulated and how long they remain in circulation, even along with aspect ratio where high aspect ratio provides longer circulation type and allows better targeting to the desired area [45–47].
- The surface charge, where positively charged particles show greater toxicity [48].
- The stiffness of the material where soft matter is much more tolerated than the hard one, suggesting hydrogels and polymer particles with embedded magnetic material to be more compatible [49].

Based on this knowledge, although very promising, each magnetic delivery system should be evaluated in and of itself before being further developed as drug delivery vehicles, contrast agents, or other potential biomedical applications.

MAGNETIC DRUG TARGETING

Therapeutic drugs are mainly administered intravenously; thus these agents do not specifically concentrate to the area of interest and instead disperse throughout the whole body causing enhanced toxicity with several side effects [50]. The essential disadvantage of most cancer treatment methods, especially of chemotherapy, is the affection of healthy cells along with the primary target [23]. Additionally the majority of traditional chemotherapeutic agents have poor aqueous solubility, selectivity, and pharmacokinetic variability [51]. Site-directed drug targeting is a possible antitumor treatment that can be applied to direct and concentrate the drug locally or regionally at the tumor site by using nanoparticles [19,52,53]. The conjugation of drug can be applied directly onto the surface of the nanoparticles or both entities can be encapsulated within a coating shell from which latter possesses the advantage of higher entrapment efficiency and stability [54]. Following the targeting of the carrier that is loaded with the drug, the corresponding release is performed in the vicinity of tissue of interest or target cells by the change in pH [55–58], temperature [56,59,60], or as a consequence of enzymatic activity [61,62], etc. [23,63]. The core benefit of this method is the possibility of designing specific drug delivery systems by considering the properties of the tumor environment such that the anticancer drugs may be targeted directly to the tumor site and affect only the cancer cells leading to the reduction or total elimination of the side effects [23,50,64]. Moreover, it is possible to deal with lower and sufficient doses of the cytotoxic compounds within the body [24].

The administration of designed drug delivery systems is still investigated from which intravenous (iv) administration and intraarterial (ia) administration are the prominent methods [65]. The advantage of intraarterial administration over conventional intravenous administration is the injection of drug-loaded nanocarrier in the vicinity of the tumor consequently preventing the uptake of nanoparticles by the reticuloendothelial system (RES) [66] before being functional at the tumor site [67]. However, nothing more than solitary tumors can be addressed by the intraarterial administration, which complicates its use in spreading tumors after metastasis.

After the administration, the targeting of drugs by using carriers can be accomplished by either passive or active targeting. For the passive targeting, the nanoparticles make use of the specific properties of the tumor environment and penetrate into the tumor by leaky vascularization as a result of enhanced permeability and retention (EPR) effect [68,69] while healthy tissues possess a barrier of endothelial cells that prevent nanoparticle extravasation [70]. This method of targeting has limited applications as a result of inefficient diffusion of drugs, nonuniform nature of tumor vessels, and the lack of EPR effect in certain types of tumors [68]. On the other hand, active targeting is based on guiding the carriers to the tumor site by the help of specific ligands that have affinity solely on the cells in question [69]. In order to target a drug conjugated nanoparticle to a desired site in the body by active targeting, often folic acid or an antibody is attached to the surface of the nanocarrier which contains the drug. Once in the system, the nanocarrier is designed for preferential binding at the tumor site due to the presence of folic acid or antibody. Although this route is frequently encountered in the literature, it has limitations in vivo due to low binding affinity.

Finally, the drug carrier should be eliminated from the system with minimum side effects following the release of the drug by metabolic activities and excretion [54].

As an alternative, externally applied magnetic field driven active drug targeting, or briefly MDT, is one of the novel route which improves the localization of the drug that is attached to an MNP by using an external magnetic field consequently concentrating the drug at the tumor site [19,52,71–73] (Fig. 6.2). Generally the magnetic carriers are composed of iron oxide cores for which magnetite is the most preferred one due to its convenient magnetic properties and a biocompatible shell or matrix for the encapsulation of therapeutic drugs. These magnetic cores not only allow for the magnetic field targeting of the entity but also bring the possibility of detection in an MRI scan [74]. Although the magnetic particles that are forming the magnetic core of the drug delivery system are required to have strong magnetic properties for convenient manipulation in the presence of an external magnetic field, it is necessary to prevent clogging of small

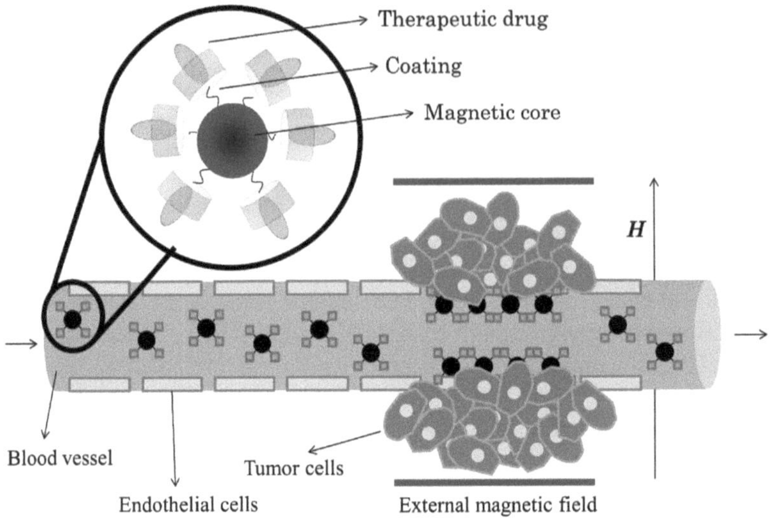

FIGURE 6.2

Targeting of magnetic drug delivery system to the tumor site by using an external magnetic field.

capillaries due to coagulation which consequently impel the application of superparamagnetic particles rather than ferrimagnetic ones having remnant magnetization even in the absence of a magnetic field [22]. Additionally, for the optimum resistance to the renal clearance, particles having sizes larger than 10 nm are reported to be useful in targeted drug delivery for increased blood circulation time after intravenous administration [75].

Similar to many other applications, there are several factors to take into consideration such as the size of MNPs (magnetic properties), characteristics of the applied magnetic field [76], binding efficiency of the drug, and physiological factors [38] when designing an effective MDT system. As in the case of many other applications, a suitable coating or a matrix is essential for the stability of magnetic particles in order to prevent agglomeration which may cause embolization of capillary vessels [23] (Fig. 6.3). This outer shell or matrix not only enhances the functionality of magnetic particles by the encapsulation of the drug and/or binding molecules but also delays the uptake by the mononuclear phagocyte system [77]. For the surface functionalization of magnetic particles that are used in magnetic drug delivery systems, several surface active agents and polymers are reported up to date such as PVA [78], PEG [79,80], chitosan [81,82], PEI [83], and starch [6,20,71,84]. These entities can further be functionalized with antibodies [85], folic acid [86], aptamers [87], etc. [6] for enhanced binding capabilities. Liposomes [6,88,89], as in the case of magnetic hyperthermia, dextran [6,60], and silica [90–92] shells may also be used for the encapsulation of

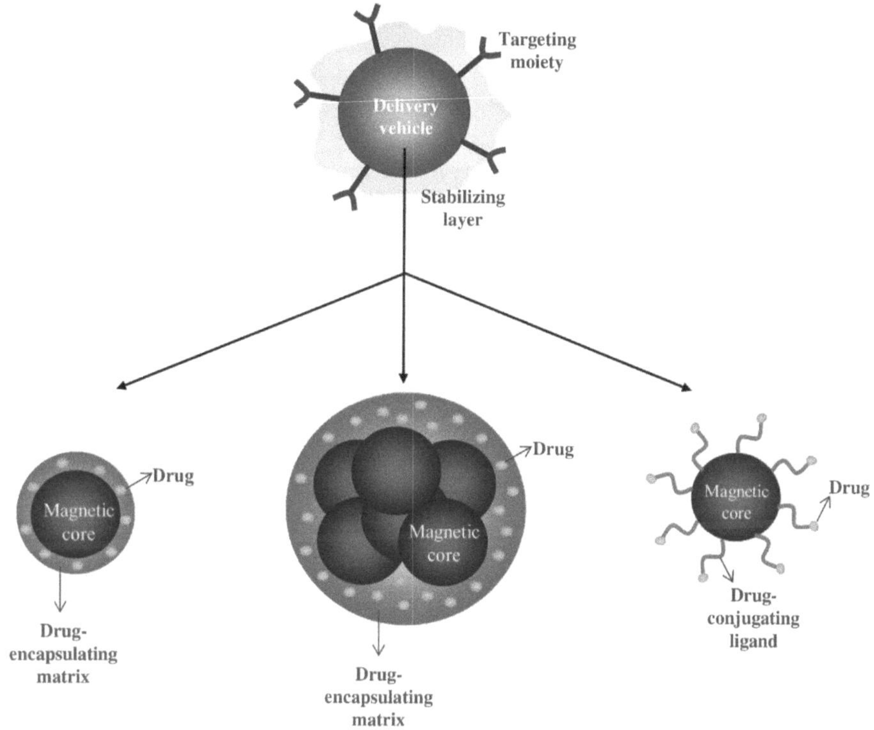

FIGURE 6.3

Schematics of different drug delivery systems containing magnetic nanoparticles.

magnetic particles and the therapeutic agents. However, it should be noted that any and every additional surface attachment have detractive effect over the magnetic properties of the core [93].

Although iron oxides are frequently encountered in magnetic drug delivery, several other metals are also reported. These metals can either be incorporated into the crystal structure or used to obtain a core–shell structure. For example, cobalt nanoparticles are incorporated in order to enhance the magnetic properties of the carrier. On the other hand, gold can be used for the direct functionalization of iron oxide surfaces to facilitate certain binding processes. Although the existence of gold on the surface of the magnetic core reduces the magnetic properties slightly, it has shown great improvement in surface binding of some antibodies. Additionally, gold surfaces are facilely functionalized therefore, gold-coated core–shell iron oxide nanoparticles are widely employed in magnetic drug delivery.

Actually, the concept of using magnetic particles for targeting dates back to couple of decades. For instance, Senyei et al. encapsulated doxorubicin in an albumin matrix containing magnetite particles and targeted these microspheres at a specific area in an in vitro analogue of the human circulatory system by using an applied magnetic field in 1978 [21]. On the other hand, Mosbach and Schroder reported in 1979 the localization of albumin-coupled magnetic starch microspheres in the ear of rabbits to which an external magnetic field (0.7 T) is applied after injection to the opposite ear after 10 min [94]. One of the leading preclinical studies for the application of MDT was reported by Lubbe et al. where an amino sugar containing positively charged anthracycline drug epirubicin is chemically bound to MNPs functionalized with anionic phosphate groups and concentrated locally in tumors by means of an external magnetic field [52]. This study paved the path for several scientists to use MNPs in drug delivery systems some of which are summarized in Table 6.1.

For instance, Chen et al. encapsulated doxorubicin-conjugated magnetite nanoparticles within a porous silica shell and further attached a layer of PEG onto the matrix. It is reported that DOX release is performed through the porous silica matrix by a diffusion-controlled process and the existence of PEG at the outer shell prevent recognition by RES thus increase the duration of DOX administration at the targeted site [90].

In another study, Singh et al. proposed an alternative system which is capable of targeting both hydrophobic (paclitaxel, rapamycin) and hydrophilic (carboplatin) drugs where superparamagnetic magnetite nanoparticles are embedded in a polylactide-*co*-glycolide matrix (PLGA-MNPs). Additionally a targeting ligand Herceptin was conjugated in order to enhance the active targeting properties. Their results illustrated the encapsulation of approximately 82% of hydrophobic and 47% of hydrophilic drugs within PLGA-MNPs followed by a successful 85% release after 3 weeks. In addition, Herceptin conjugation on the surface of PLGA-MNPs was shown to enhance the selectivity while the overall targeting lead to a cytotoxic activity in MCF-7 and PANC-1 cells in in vitro studies [95].

It is also possible to use a dual-drug delivery systems, which assemble different therapeutic drugs of diverse features such as antitumor and antiangiocardiopathy activities in the same matrix [96].

In another study a specific peptide, chlorotoxin (CTX) which is used in combination with iodine for the targeting of radiation to tumor cells is conjugated with iron oxide nanoparticles due to its subfunction in affinity for specific types of tumors (NP-CTX) [99]. The drug delivery system that contains the chemotherapeutic drug methotrexate (MTX) is shown an enhanced accumulation in glioma and medulloblastoma cells in in vitro studies for MRI applications. In vivo, a retention time of at least 2 weeks for the chemotherapeutic agent is also demonstrated for a 9L xenograft tumor-bearing mice [97].

In recent years, the application of two-dimensional graphene and its corresponding composite nanomaterials rather than conventional systems in MDT was reported by some authors for their excellent mechanical, thermal, and structural properties. For instance, Ma et al. illustrated the promising use of PEG-functionalized graphene oxide-superparamagnetic iron oxide hybrid nanocomposite (GO-IONP-PEG) in both MDT and MRI. The GO-IONP-PEG was shown to

Table 6.1 Summary of Selected In Vivo and In Vitro Studies Involving Magnetic Drug Delivery Systems

Magnetic Core	Therapeutic Drug	Coating/Shell	Target Cells' Tumor Type
[55] $MnFe_2O_4$	Doxorubicin (DOX)	Herceptin-modified Py-PEG-COOH pyrenyl-PEG	NIH3T6.7 Breast cancer
[56] Fe_3O_4	DOX	P(NIPAM-*co*-MAA)-coated mesoporous silica shell	HeLa cervical carcinoma
[58] Fe_3O_4@SiO_2	DOX	mPEG-poly (L-Asparagine)	MDA-MB-231 breast tumor
[63] $CoFe_2O_4$	Paclitaxel (PTX)	$BaTiO_3$	SKOV-3 human ovarian carcinoma
[71] Iron oxide	Mitoxantrone (MTX)	Starch	VX2 squamous cell carcinoma
[72] Iron oxide	Epirubicin	Anhydroglucose polymers	Metastatic breast cancer, chondrosarcoma, squamous cell carcinoma
[73] Iron oxide	MTX	Starch	squamous cell carcinoma
[79] Fe_3O_4	MTX	PEG	9L-glioma
[80] Fe_3O_4	DOX	Oleic acid – PEG	MCF-7 breast cancer
[20] Fe_3O_4	MTX	Starch	VX2 squamous cell carcinoma
[84] Fe_3O_4	MTX	Starch	VX2 squamous cell carcinoma
[87] Iron oxide	DOX	A10 RNA aptamer and PEG	Prostate cancer cells
[89] Fe_3O_4	PTX	Liposomes	EMT-6 breast cancer
[90] Fe_3O_4	DOX	PEG-functionalized porous silica shell	MCF-7 and HeLa cells
[91] Fe_2O_4@SiO_2	Docetaxel (DOC)	Folic acid and PEG	FR(+) HeLa cells and FR(−) MCF-7 cells
[92] Fe_3O_4@SiO_2	DOX	PEG	MCF-7 breast cancer cells
[95] Fe_3O_4	PTX, Rapamycin, carboplatin	Herceptin-modified PLGA	MCF-7 and PANC-1
[96] Fe_3O_4	DOX, verapamil (VER)	Cyclo(ArgGly-Asp-D-Phe-Lys) conjugated PLGA	S-180 murine sarcoma cells
[97] Fe_3O_4	Methotrexate	Chlorotoxin and PEG	9L-cells
[98] Iron oxide	DOX	Graphene oxide and PEG	4T1 breast cancer cells

be effective as T2 contrast agent in in vivo MRI of tumor-bearing mice as well as in in vitro direct targeting of DOX to the murine breast cancer 4T1 cells [98].

In another study, Tietze et al. investigated the effect of administration type over in vivo biodistribution of Mitoxantrone (MTO) after MDT at the targeted tumor tissue which is implanted in the limbs of rabbits. In order to analyze the biodistribution, MTO was both administered alone and with superparamagnetic particles intravenously in the presence and absence of an external magnetic field. Additionally, MTO-loaded superparamagnetic particles were also administered by intraarterial injection with an external magnetic field. Their results indicated that MDT after intraarterial injection resulted in a considerable accumulation (four-fold) of MTO at the tumor site in comparison to liver and kidneys while intravenous administration led to the accumulation of MTO in kidneys (>75%) decreasing the doses below 1% at the tumor site [67]. In line with this study, tumor entrapment of PEI-modified MNPs were resulted in an extraordinary 30-fold increase after intraarterial administration followed by magnetic targeting when compared with intravenous administration [83].

Along with its advantages, MDT has also some drawbacks which affect its clinical applications. One of them is again the systemic distribution of therapeutic drugs. These drugs are targeted to the tumor site by magnetic cores (particles) under the influence of a high-gradient magnetic field, in the absence those do not have any retainer force for stabilization at the target site which still causes some systemic distribution [24]. On the other hand, magnetic field intensity may not be sufficient and limiting to affect sites rather than near-surface targets and also cause accumulation of magnetic particles at another location where the magnetic field gradient is relatively more perceived [24,100].

In principle, it can be said that in the design of MNPs in drug delivery, particles with high magnetization cores that already exhibit low cytotoxicity, should be individually coated or embedded in a matrix that can encapsulate the drug, or the drug should be conjugated to the particle, after which the drug-encapsulating entity can be made more biocompatible and their retention in circulation can be enhanced by the addition of some stabilizers and they can be further directed to the tumor site by being functionalized by targeting moieties in addition to being targeted by the application of an external magnetic field. The ability of being targeted bimodally, makes MNPs rise in novel drug delivery applications.

HYPERTHERMIA

Hyperthermia is an alternative therapeutic method for the treatment of cancer by procuring heat at the tumor site. In hyperthermia, the cancer cells are affected adversely by enhancing the temperature of a region of the body locally (41–46°C) resulting in the partial or complete destruction of the tumor by either a programmable (apoptosis) or a premature death (necrosis) of malignant cells as a consequence of mechanisms such as protein denaturation, protein folding, and DNA cross-linking [14,101–105]. Hyperthermia can be applied locally, regionally, or to whole body considering the conditions and dissemination of cancer [106]. Magnetic hyperthermia is applied by using suitable MNPs where the reorientation of the magnetization of MNPs is achieved in the presence of an external magnetic field. This continuous reversal of magnetization process causes losses and heat generation thus a local temperature increase which results in death of cancer cells or increases susceptibility for the complementary treatment methods such as surgery, chemotherapy, radiation therapy, and gene therapy [107,108]. Previous studies showed that there is a promising progression in the treatment of cancer as a result of the combination of these therapeutic methods [109–112]. There are several advantages of using MNPs in hyperthermia such as targeting the particles to the tumor site by the application of an external field which can then be applied as AMF in order to stimulate the heating. It is also possible to encapsulate drugs to magnetic carriers to facilitate effective drug delivery systems as well as enhancing the treatment efficiency of hyperthermia. The effective use of MNPs in magnetic hyperthermia depends on the intensity and the frequency of the applied magnetic field, exposure time as well as the concentration, size, morphology, and magnetic properties of the nanoparticles [8,113–115]. There exists several in vitro and in vivo studies related with magnetic hyperthermia are performed at various conditions from the pioneering studies to this date [14,15,17,107,116–121]. Essentially, the magnetic field frequencies and intensities are indicated to be in the range of $f = 0.05-1.2$ MHz and $H = 0-15$ kA/m while the concentration of the MNPs is about 5–10 mg/cm^3 of tumor [18,122]. The heating ability of MNPs in magnetic hyperthermia is given by the specific absorption rate (SAR), which is the heating power obtained per unit mass of particles which is also referred in some studies as the specific loss power (SLP) or the specific power loss. SAR is given by

$$SAR = C\frac{\Delta T}{\Delta t}$$

where C is the specific heat capacity of the sample and $\Delta T/\Delta t$ defines the temperature enhancement per unit time [120]. The heating ability of the MNPs is proposed to be due to the combination of Néel relaxations and Brownian relaxation processes, which are the physical rotations of particles within the nanofluid and the rotation of magnetic moments with in the particles, respectively [18,107,123]. An intrinsic parameter, the intrinsic loss power is also recommended as it excludes the extrinsic factors such as the frequency and the intensity of the applied magnetic field [124]. Consequently, for the optimum application of MNPs in magnetic hyperthermia, it is desired to achieve an effective temperature increase by using minimum amount of MNPs for which the SLP must be sufficiently high. One of the most challenging problems of magnetic hyperthermia is the difficulty in selective targeting or homogeneous distribution of MNPs at the tumor site. An alternative approach is to target surface functional MNPs to the tumor site using an external magnetic field in order to increase the local concentration [125]. In addition, functionalized MNPs may be modified in such a way that a chemotherapeutic drug is also incorporated to the entity for the direct administration to the tumor site [126,127].

Although there are quite different magnetic materials proposed for the application of magnetic hyperthermia, overall inclination is the use of biocompatible (nontoxic) magnetic iron oxides, magnetite (Fe_3O_4), and maghemite (γ-Fe_2O_3) [105]. For magnetic hyperthermia applications initially the use of large multidomain particles or smaller superparamagnetic particles was preferred. However, it was later on shown that the most advanced SLP may be achieved by the use of magnetic particles in the transition range from superparamagnetic to ferrimagnetic properties where heating arises as a result of hysteresis losses due to magnetic domain wall displacements or Neél relaxations in the presence of an AC magnetic field [16,120]. There are also contrary reports stating that superparamagnetic nanoparticles give higher SAR [120]. On the other hand, the maximum SLP (1 kW/g) is revealed for bacterial magnetosomes where the particle size is around 30–35 nm [128]. As the isolation of these perfect uniform MNPs from the bacteria is compelling, studies focus on obtaining similar single-domain MNPs that are readily stabilized in suspensions for magnetic hyperthermia applications [32]. Although obtaining the maximum amount of SAR with MNPs is an essence, it is a challenge for in vitro and clinical studies to target these specific particles and obtain a uniform distribution throughout the tumor. A review of experimentally determined, both in vivo and in vitro, SAR/SLP for different nanoparticle systems and magnetic field properties is given in Table 6.2.

For a couple of decades, there has been countless studies on magnetic hyperthermia both in vitro and in vivo all of which pave the way of using MNPs in human patients [138]. For the determination of suitable conditions for the application of magnetic hyperthermia in vivo or in clinical studies, it is necessary to analyze the performance of the method in vitro where SLP of MNPs is attempted to be optimized by the variation of parameters such as concentration of MNPs, specific properties of the applied magnetic field, and tested cell types, etc by applying induction heat studies and calorimetric methods [15]. At the same time, for these studies several surface active agents such as dextran [139], lauric acid [140], silane [132], cationic liposomes [141], oleic acid [133], folic acid [130], and citric acid [142] are incorporated onto the surface of MNPs in order to improve the colloidal stability as magnetic particles tend to agglomerate and enhance the surface functionality. It was shown in vitro that cellular uptake of superparamagnetic magnetite nanoparticles that were functionalized with dextran or aminosilane groups in glioblastoma cells was substantially higher as in normal cells which shows the possibility of active targeting without a tagging ligand [132]. Additionally, it was shown that reproduced cancer cells concurrently contain MNPs that were taken by predecessor cells such that consequent MFH is possible without additional targeting of MNPs [14]. On the contrary, another report showed that nanoparticles had significantly taken up by normal tissue cells as

Table 6.2 Summary of SAR/SLP for Different Magnetic Nanoparticle Systems

Magnetic Nanoparticle	Core Size (Hydrodynamic Size) (nm)	Coating	*H* (kA/m)	*f* (kHz)	SAR/SLP (W/g)
[115] Fe_3O_4	11.2	Pluronic F127	24.5	400	447
[125] Fe_3O_4	10.3	Chitosan	–	150	64.6
[125] Fe_3O_4	10.3	Starch	–	150	62.3
[125] Fe_3O_4	10.3	None	–	150	69.8
[129] Fe_2O_3	23(160)	Carboxydextran	11	410	400
[130] Fe_3O_4	9.3(98.9)	Folic Acid - PEG	8	230	248
[131] Fe_3O_4	15(217)	Dextran	12.7	250	46
[132] Fe_3O_4	3.3(50–70)	Dextran	13.2	520	120
[132] Fe_3O_4	13.1(17)	Aminosilan	13.2	520	146
[133] Fe_3O_4	10	None	26	265	34.6
[133] Fe_3O_4	10	Oleic acid	26	265	40.8
[134] Fe_3O_4	19	None	24	700	2277
[134] Fe_3O_4	19	None	22	325	1000
[135] Fe_3O_4	8(22)	Cystene	8	230	156
[136] Fe_3O_4	20	None	26.6	265	38.4
[136] Fe_3O_4	6	Oleic acid	26.6	265	28.3
[136] Fe_3O_4	10	PEG	26.6	265	33.5
[137] Fe_3O_4	12.1(28)	BSA	6.3	400	36
[137] Fe_3O_4	10.4	Citrate	6.3	400	5.2

well as tumor cells after iv administration of magnetite nanoparticles into a tail vein of rats having mammary tumors, indicating the importance of ligand labeling [117]. In a similar manner, the effect of superparamagnetic magnetite nanoparticles in a dextran shell over the death rate of dendritic cells in the presence of an alternating magnetic field was investigated by Asin et al. for determining the optimum particle concentration as well as exposure time and field amplitude. Their results specifically point out the necessity for the proper choice of magnetic field properties for magnetic hyperthermia applications [131]. Ghosh et al. represented that oleic acid and PEG-coated superparamagnetic magnetite nanoparticles achieve up to 65% reduction in the human breast cancer cells (MCF-7) under heating conditions [136]. Guardia et al. also reported the hyperthermia performance of PEG-functionalized MNPs in the size range between 13 and 40 nm as a function of magnetic field intensity and frequency. Their results indicated one of the highest values of SAR (2452 W/g) at 520 kHz and 29 kA/m and almost 50% mortality of KB cancer cells at 43°C after 1 h of treatment in vitro [134]. Similarly, heating ability that is comparable to that of magnetosomes are obtained for cysteine-coated superparamagnetic magnetite nanoparticles [135].

Samanta et al. reported that protein (BSA) passivated superparamagnetic magnetite nanoparticles cause total destruction of HeLa cells after 45 min exposure to an alternating magnetic field (400 kHz and 6.3 kA/m) [137]. Differently, Saavedra et al. proposed the advantage of using unique coating agents that have ordered pore structures on the multimodal therapies. Results indicated the internalization of maghemite nanoparticles within a mesa-porous silica matrix by human lung carcinoma cells. It was further shown that magnetic microspheres may be used for heat treatments leading to a decrease in the cell viability by AMF exposure [143]. In vivo, essentially, the application of magnetic nanofluids in hyperthermia applications is performed via injection into the tumor site or by iv administration [144]. One of the most important necessity for the efficient application of magnetic hyperthermia is obtaining adequate particle concentration and uniform distribution at the tumor site which is still a challenge to overcome [145]. The first application of magnetic particles for hyperthermia in vivo was in 1957 when Gilchrist et al. performed selective heating of lymph nodes by using magnetite [116]. This study was followed by countless experimental studies at various conditions. Jordan et al, who illustrated the heating of superparamagnetic dextran coated magnetite nanoparticles that were injected into mammary carcinomas of mice [119]. Magnetic cationic liposomes (magnetoliposomes) were also injected into rat glioma and shown to be potentially effective by both destroying the tumor cells and inducing an immune response [146]. In another study, for hamster osteosarcoma, these liposomes were shown to regress the tumor almost 100% in the presence of AMF where no relapse was observed after 3 months [147]. Hayashi et al. addressed the low accumulation in tumors due to the size of superparamagnetic nanoparticles and synthesized clustered folic acid and PEG-coated MNPs to prevent leakage from fenestrated capillaries in tumors and enhance SAR of individual particles. Their results indicated a local accumulation of MNPs in tumor tissues of mice and a 6°C difference in the temperature of tumor with respect to the surrounding tissues at 20 min after treatment [130]. In 2007 the first application of magnetic hyperthermia for the treatment of glioblastoma was successfully presented by Jordan et al. in human patients by direct injection of superparamagnetic magnetite nanoparticles (15 nm and 112 mg/mL) to multiple tumor sites that were exposed to an alternating magnetic field up to 13.5 kA/m at 100 kHz [138]. On the other hand, Lee et al. performed in vivo hyperthermia treatment with cancer cells (U87MG) and compared the effect of chemotherapeutic drug doxorubicin with magnetic thermal induction by nanoparticles at the same concentration over tumor destruction. After 18 days the tumor was completely eliminated for the group where magnetic hyperthermia was applied with core–shell nanoparticles. On the other hand, initially regressed tumor size for the group treated with doxorubicin come up with four times its original figure after 26 days [148]. Finally, Balivada et al. reported that even microgram amounts of core–shell porphyrn-labeled dopamine-anchored Fe/Fe_3O_4 nanoparticles is sufficient to cause an antitumor effect on melanoma in mice with relatively short repetitive exposure time (10 min) [149].

As summarized in this section, hyperthermia is a unique application of MNPs, as the local increase in temperature is achieved upon application of an external magnetic field. Due to the need of particles with high specific adsorption rates, choice of magnetic core is very important. Specific binding of magnetite to the tumor site achieved by targeting is favored to homogeneously distribute the particles around the tumor to obtain a homogeneous heating profile. Similar drug targeting strategies can be utilized as summarized in "Magnetic Drug Targeting" section to simultaneously delivery drugs to the tumor site to enhance the efficiency of hyperthermia. The predominantly in vitro success of the materials at this point, only points to their potential in vivo applications where by better distribution and retention of particles at the tumor site, clinical studies will soon flourish.

MAGNETIC RESONANCE IMAGING

MRI is a widely used technique in the imaging of soft tissue, which in principle is based on Nuclear Magnetic Resonance (NMR). When placed in a magnetic field, different atomic nuclei absorb and emit electromagnetic radiation which is in the radio frequency. Hydrogen atoms, which are most abundant in living organisms predominantly in the form of water or hydrocarbon (i.e., fat or sugar), are often used to generate a detectable signal. Due to the presence of hydrogen atoms in the soft tissue, MRI technique is often used to detect tumors, hemorrhages, etc.

There are mainly two types of relaxations, T1 which is longitudinal is the time for the nucleus to relax back to its thermal equilibrium after being probed by electromagnetic radiation. The other type of relaxation is T2 which is transverse and refers to loss of phase coherence.

In obtaining images, different contrast modes can be used depending on the target tissue. In some cases, more information can be revealed using either T1- or T2-weighed images, whereas mostly a combination serves best. Often a contrasting agent is administered before imaging to shorten the relaxation times and obtain better images. Gadolinium-based contrast agents are most widely used [150–152], although some superparamagnetic iron oxides are also FDA approved and are currently being used. Gadolinium-based contrast agents which are paramagnetic shorten the T1 relaxation, whereas MNPs lower the T2 signal, enhancing the contrast, resulting in higher resolution images. As we are focusing on MNPs in this chapter, extensive work with gadolinium-based contrast agents will not be discussed.

Enhancement of contrast due to the lowering of the T2 signal is directly proportional with the magnetic moment of the material. Magnetite (Fe_3O_4) is the naturally occurring iron oxide with the highest magnetization. It is also possible to increase the magnetic moment by doping Fe_3O_4 with other transition elements that are highly magnetic such as Co, Mn, and Ni [153–156] to obtain better contrast agents. For the same purpose, it is also possible to synthesize iron alloys such as FeCo [157,158] or FePt [159,160] to be used for MRI. However, a major issue with particles containing metals other than iron is the cytotoxicity [161,162]. Use of naturally occurring iron-based nanoparticles such as ferritin as an MRI agent is also suggested especially due to its excellent biocompatibility [163]. Carbon nanotubes are also exploited as contrast agents where their surfaces are either decorated with maghemite [164] or magnetite [165,166] nanoparticles; or they are filled with a ferromagnetic material [167]. The main focus of this section will be on the use of iron oxide MNPs as contrast agents for MRI, and other magnetic materials will be excluded.

Clinical studies have been carried out using FDA-approved MNPs. There are very specific contrast agents used to differentiate tumors. Lymph node sizes should be accurately determined to differentiate whether the lymph node is metastatic or not. FDA-approved ferumoxtran (a.k.a. Combidex) which consist of monocrystalline superparamagnetic nanoparticles with a dense dextran derivative coating was used in a Phase III clinical trial on 220 patients where persistently high T2* signal intensities were achieved [168,169]. In another study, same particles were shown to demonstrate high sensitivity and specificity in distinguishing metastatic lymph nodes in patients with renal cell cancer [169]. FDA-approved Resovist which consist of superparamagnetic magnetite particles coated with carboxydextran of about 60 nm is predominantly used for the detection and characterization of especially small focal liver lesions [170]. To cure type 1 diabetes, Islets of Langerhans transplantation has been suggested; however a noninvasive assessment such as MRI to follow the retention of islets is required [171]. In a clinical study, Resovist-labeled islets have been transplanted to patients where early detection of islet loss was monitored [171]. In a Phase III trial, same particles were employed for the visualization of myocardial infarct zone; however, no improvement compared to gadolinum-based contrast agents was achieved [172]. Both Resovist and Feridex (small superparamagnetic iron oxide particles with low-molecular-weight dextran coating of about 150 nm size) were shown to be used for the evaluation of macrophage activities as well as imaging liver [173]. Ultrasmall iron oxide nanoparticles (Sinerem, stabilized with dextran and sodium citrate) were proven useful to visualize normal sized pelvic lymph nodes [174]. Clariscan, which consists of small MNPs with PEG coating, was developed as a gastrointestinal contrast agent, also tried for coronary angiography with reasonably good results [175]. Although once approved, neither of the abovementioned products is now available in the market due to limited applications or safety concerns. On the other hand, GastoMARK (a.k.a. Lumirem) which is a superparamagnetic iron oxide nanoparticle aggregates of 300 nm size with dextran coating is currently used as contrast agent in intestinal movements and pancreas and is well received by patients [176,177].

Despite some practical problems in the use of superparamagnetic iron oxide nanoparticles as MRI contrast agents, a vast number of particles are currently being developed and tested. Even bare magnetite nanoparticles are shown to lower the T2 signal [178], however due to cytotoxicity related to bare particles and lack of colloidal stability resulting in aggregation, brings the necessity to generate particles with different coatings. In order to be applicable as MRI contrast agents, particles should remain in circulation a certain amount of time before being cleared. One of the criteria to ensure that is to synthesize particles that fall within a size range. Particles that are too small (<5 nm) are quickly removed by the kidney, whereas particles of about 20 nm are cleared by the liver. Use of larger particles is even more challenging as they are uptaken by the immune system. It should be noted that the abovementioned sizes are hydrodynamic radia, which refer to the magnetic core and the stabilizing layer around the particles.

One of the simpler anionic magnetite nanoparticles with an oleic acid double layer were presented as potential MRI contrast agents [179]. Ultrasmall superparamagnetic iron oxide nanoparticles, when synthesized with narrow size distribution and high polymer coverage with low-molecular-weight polyacrylic acid (PAA, MW 5100 Da), exhibited high colloidal stability and dual T1-T2 enhancement [180]. Although endocytosis (cell uptake) of anionic particles is presented [181], cationic nanoparticles, providing better electrostatic interaction with the negatively charged cells, are better uptaken [182]. Ethyl amine–functionalized ultrasmall SPIONs are shown in vitro and in vivo to be good contrast agents [183]. Additionally, octopod magnetite nanoparticles are shown to be more effective as a T2 contrast agents for in vivo imaging than conventional spherical ones, indicating the morphology dependence of T2 relaxivity (Fig. 6.4) [184]. Despite their name as ultrasmall particle, the size of these particles still exceeds the lower size limit for kidney clearance and they are shown to accumulate in the liver, which is the preferred secretion route out of the body for MRI applications.

However the amount of time the particles remain in circulation is also closely related to their coating. To enhance biocompatibility and colloidal stability, as in the case of liposomes, PEGylation of particle surface is a common strategy also for the development of contrast agents. Due to the presence of PEG, particle circulation time in blood increases, making in vivo imaging more viable [185]. Cellular uptake of PEG-modified magnetite particles were achieved using poly-L-lysine transfection agent (cationic, therefore favorable due to induce cellular uptake), exhibiting promising in vivo MRI contrast enhancement [186]. Small superparamagnetic iron oxides when embedded in a shell of cross-linked PEG and polyamino acids (a.k.a. polyion complex vesicles) are shown to perform several times better as MRI contrast agents and detect small tumors [187]. PEG can also be used as an outer corona for drug-loaded particles in their inner shell for simultaneous drug delivery and high T2 relaxivity values [188]. Hydrophobic magnetite nanoparticles embedded in the bilayer of phospholipid liposomes [189] are also developed to obtain better contrast agents, taking advantage of the excellent biocompatibility and stability of liposomes while benefitting from the lipid-enveloped magnetic core.

For highly sensitive MR imaging, prolonged retention in the blood stream as well as accumulation in the tumor area are required. A significant enhancement in

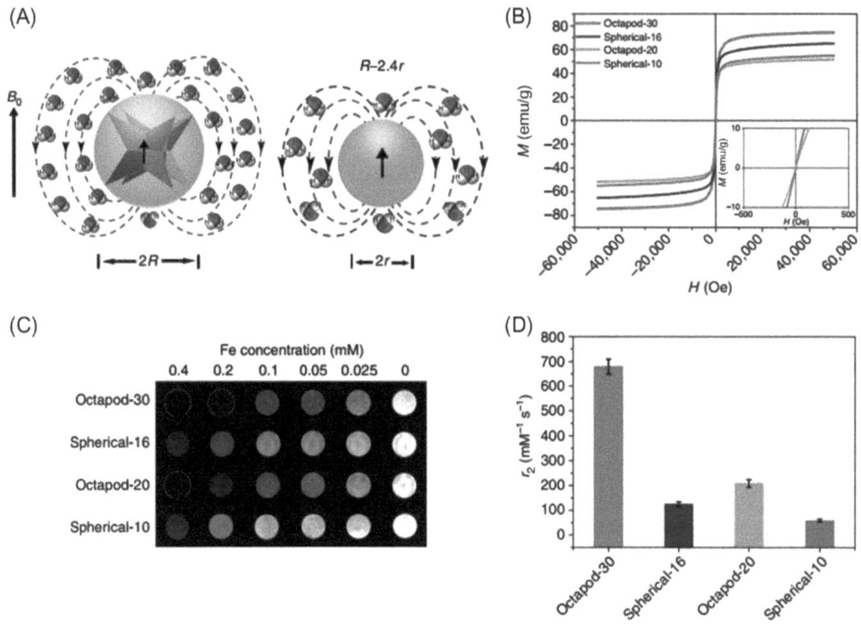

FIGURE 6.4

The effect of octopod iron oxide nanoparticles over the MR contrast. (A) Ball models and effective volume, (B) magnetization curves, (C) T2-weighted MR images, and (D) r_2 values of octopod and spherical iron oxide nanoparticles.

magnetic resonance contrast was obtained due to a dense iron oxide core and large cluster size by covalently conjugating β-cyclodextrin to superparamagnetic particles and hosting PEG in its cavity where the cluster size was controlled by particle/PEG ratio [190].

Multicore MNPs of magnetite and maghemite with a diethylaminoethyl-dextran shell were embedded in a gel of petroleum oil and styrenic copolymers at various concentrations and these materials were tested for their suitability for XCT and MRI. Although limited due to the nature of the gel, these materials were acceptable T2 enhancers for MRI [191].

All the abovementioned particles reach the tumor site by passive targeting. In passive targeting, nanoparticles reach and accumulate at the tumor site due to leaky vascularization. The released drug in the blood diffuses into the tumor tissue passively. This brings some inefficiency in targeting, therefore efficient active targeting where MRI agent nanoparticles are coupled with targeting moieties becomes the purpose of cutting edge research.

Folate receptor protein is a protein that is on the surface of cancer cells, therefore is overly expressed at the tumor site. Folate is a vitamin that strongly binds to this protein which makes it a common targeting moiety to accumulate the nanoparticles at the tumor site. Folic acid ligands are conjugated onto the $-NH_2$ groups of magnetite nanoparticles obtaining colloidally stable particles with a core size of 8 nm and a hydrodynamic diameter of about 26 nm. As a result of preferential binding on the tumor cells, the MRI signal is shown to be enhanced, generating a higher contrast between tumor and normal tissue [192]. Magnetite-loaded chitosan particles, when compared with its folic acid counterpart were shown to accumulate better at the tumor site and can be used as an organ-specific MRI contrast agent [193]; however even in the absence of folic acid, magnetite-chitosan particles were also shown to significantly decrease T2 intensity [194]. Carboxymethyl dextran–modified magnetite particles with folic acid were successfully shown to enhance MRI contrast in the detection of KB cells [195]. Magnetite-encapsulated silica particles were conjugated with folic acid and despite the low magnetization values, were shown to be potential candidates as MRI contrast agents [196]. Magnetite nanoparticle surfaces were modified with folate-PEG-caffeic acid–modified magnetite nanoparticles were preferentially taken by a folate receptor overexpressed cancerous cells and the relaxation times increased with particles concentration [197].

Other targeting moieties, such as antibodies can also be conjugated to enhance accumulation at the target site to be used as MRI agents. Specifically image breast cancer tumors, human epidermal growth factor receptor 2 (anti-HER2) antibody was conjugated onto silica-magnetite core–shell particle surfaces and successful binding resulting in MRI detection with better contrast was supported with in vitro and in vivo studies [198].

Liver targeting of MNPs was achieved attaching a liver targeting function mebrofenin on silica-coated magnetite nanoparticles. Effective T2 signal decrease proved these particles as liver targeting contrast agents [199].

Early diagnosis of one of the more aggressive cancers, pancreatic cancer is of utmost importance. Galectin-1 is expressed in pancreatic cancer cells and its natural ligand glycosylated peptides can be attached onto maghemite-embedded recombinant human serum albumin nanoparticles as a targeting moiety. Although MRI contrast enhancement was achieved either way, in the absence of targeting, the particles were mostly found in the liver, whereas the uptake of magnetic particles with the targeting ligand was significant [200].

A specific targeting peptide, chlorotoxin was covalently bound to silica-magnetite core–shell particle surface due to its preferential binding to glicoma cells. As a result of more efficient uptake by the target cells, MRI contrast between glicomas and normal brain tissue was enhanced [201].

To take advantage of the photothermal effect of gold, magnetite nanoparticles were coated with gold and then functionalized with thiol-modified aptamers as the targeting moiety, where they could simultaneously serve as MR agents and used in photothermal therapy [202]. A similar construct with gold-coated magnetite nanoparticles, by the attachment of DNA strands onto the surface, self-assembly to form nanoparticle chains were induced to improve MRI detection modality [203].

MRI is not only used in the detection of tumor in the body but also used for the tracking of cells after stem cell therapies. In stem cell transplantation, noninvasive evaluation of the location and distribution of the stem cells is very important. For the MRI of stem cells, meso-2,3-dimercaptosuccinic acid (DMSA)-coated maghemite nanoparticles of around 30 nm were synthesized. Lowering of T2 signal in vitro studies were in line with successfully higher contrast obtained in vivo [204]. For a similar purpose, single strand DNA was conjugated with 1,2-dihexadecanoyl-*sn*-glycero-3-phosphoethanolamine (DPPE), which is a lipid using PEG as a linker. This lipid got inserted in a cell membrane, leaving ssDNA on the cell surface for conjugation. Once the magnetite nanoparticles surface is modified with the complementary strand, DNA hybridization allow cell labeling where MRI scanning of transplanted cells was possible up to 1 month [205]. Efficient gene delivery and its post therapy monitoring with MRI were simultaneously achieved using cationic polyethylene imine (PEI)–coated chitosan embedded micelles accommodating hydrophobic superparamagnetic iron oxide nanoparticles in their core. The plasmid was adsorbed onto the particle surface due to electrostatic interactions [206].

As seen in previous sections, MNPs have versatile applications for the detection and treatment of cancer. Some particles are developed to serve more than one function. One example to drug delivery in conjunction with MRI is the gelatin-encapsulated magnetite nanoparticles conjugating some drugs onto their surfaces [207]. Doxorubicin (DOX) is an anticancer drug that is frequently used as a modal drug in drug delivery studies as in the following. BSA shell with a PEG corona around MNPs [188], magnetite containing mesoporous silica nanoparticles [208], and superparamagnetic nanoparticles coated by an amphiphilic block copolymer are some examples of simultaneous drug delivery and contrast enhancement systems [209].

Multipurpose nanoparticles for hyperthermia, MRI, and drug delivery were developed by functionalizing magnetite with β-cyclodextrin (β-CD) and coating with a pluronic surfactant. Curcumin was used as a model drug to be delivered from the outer shell of the nanoparticle and the magnetic core provided enhancement of MRI contrast, where the overall particle exhibited better performance in hyperthermia [210]. Hyaluronic acid was used to target CD44-expressed cancer by being conjugated onto magnetite surface and PEG was added to prolong circulation time. In this comparative study better contrast was achieved in the absence of PEG, whereas both types of particles were found to be equally effective in hyperthermia applications [211].

In general, MNPs, especially those with iron oxide cores, with low cytotoxicity, have shown great potential in lowering the T2 signal, enhancing contrast in MR imaging. By enhancing the circulation time, addition of drug molecules, and targeting moieties, these particles are now further optimized to be more frequently used in clinical applications.

To summarize, iron oxide MNPs have shown great promise in cancer research, due to high SAR values which make them valuable for the shrinking or disappearance of tumors as a result of hyperthermia upon application of an alternating current, due to their magnetic core with high magnetic moment which lend these particles as contrast enhancing agents for MRI as well as facilitating their manipulation upon application of an external magnetic field. The relatively facile control over their size, allowing tunable magnetic properties make these particles to partake in versatile applications. Due to the ease of surface functionalization, MNPs can be tailored to be involved in the encapsulation and conjugation of several anticancer drugs while some surface modifications are performed for their prolonged circulation in the blood to become more efficient cancer theranostics agents. Growing research and clinical studies have shown iron oxide nanoparticles to be rather applicable in cancer diagnosis and therapy, hinting future to bring more end-user products with these particles.

REFERENCES

[1] WHO. Global cancer rates could increase by 50% to 15 million by 2020. Geneva, Switzerland: WHO; 2003.

[2] Repetto L. Greater risks of chemotherapy toxicity in elderly patients with cancer. J Support Oncol 2002;1(4 Suppl. 2):18–24.

[3] Brannon-Peppas L, Blanchette JO. Nanoparticle and targeted systems for cancer therapy. Adv Drug Deliv 2004;56(11):1649–59.

[4] Coates A, Abraham S, Kaye SB, Sowerbutts T, Frewin C, Fox R, et al. On the receiving end—patient perception of the side-effects of cancer chemotherapy. Eur J Cancer Clin Oncol 1983;19(2):203–8.

[5] Hao R, Xing R, Xu Z, Hou Y, Gao S, Sun S. Synthesis, functionalization, and biomedical applications of multifunctional magnetic nanoparticles. Adv Mater 2010;22 (25):2729–42.

[6] Veiseh O, Gunn JW, Zhang M. Design and fabrication of magnetic nanoparticles for targeted drug delivery and imaging. Adv Drug Deliv Rev 2010;62(3):284–304.

[7] Lee J-H, Huh Y-M, Jun Y-W, Seo J-W, Jang J-T, Song H-T, et al. Artificially engineered magnetic nanoparticles for ultra-sensitive molecular imaging. Nat Med 2007;13(1):95–9.

[8] Gupta AK, Gupta M. Synthesis and surface engineering of iron oxide nanoparticles for biomedical applications. Biomaterials 2005;26(18):3995–4021.

[9] Bock N, Riminucci A, Dionigi C, Russo A, Tampieri A, Landi E, et al. A novel route in bone tissue engineering: magnetic biomimetic scaffolds. Acta Biomater 2010;6 (3):786–96.

[10] Matsunaga T, Kawasaki M, Yu X, Tsujimura N, Nakamura N. Chemiluminescence enzyme immunoassay using bacterial magnetic particles. Anal Chem 1996;68 (20):3551–4.

[11] Nakamura N, Hashimoto K, Matsunaga T. Immunoassay method for the determination of immunoglobulin G using bacterial magnetic particles. Anal Chem 1991;63 (3):268–72.

[12] Kutushov MV, Kuznetsov AA, Filippov VI, Kuznetsov OA. New method of biological fluid detoxification based on magnetic adsorbents. Scientific and clinical applications of magnetic carriers. New York, NY: Plenum Press; 1997. p. 391–7.

[13] Hahn GM. Hyperthermia and cancer. New York, NY: Plenum Press; 1982.
[14] Jordan A, Scholz R, Wust P, Fähling H, Felix R. Magnetic fluid hyperthermia (MFH): cancer treatment with AC magnetic field induced excitation of biocompatible superparamagnetic nanoparticles. J Magn Magn Mater 1999;201(1):413–19.
[15] Salunkhe AB, Khot VM, Pawar S. Magnetic hyperthermia with magnetic nanoparticles: a status review. Curr Topics Med Chem 2014;14(5):572–94.
[16] Hergt R, Dutz S, Müller R, Zeisberger M. Magnetic particle hyperthermia: nanoparticle magnetism and materials development for cancer therapy. J Phys Condens Matter 2006;18(38):S2919.
[17] Hergt R, Dutz S. Magnetic particle hyperthermia—biophysical limitations of a visionary tumour therapy. J Magn Magn Mater 2007;311(1):187–92.
[18] Pankhurst QA, Connolly J, Jones S, Dobson J. Applications of magnetic nanoparticles in biomedicine. J Phys D Appl Phys 2003;36(13):R167.
[19] Lübbe AS, Alexiou C, Bergemann C. Clinical applications of magnetic drug targeting. J Surg Res 2001;95(2):200–6.
[20] Alexiou C, Jurgons R, Schmid RJ, Bergemann C, Henke J, Erhard W, et al. Magnetic drug targeting—biodistribution of the magnetic carrier and the chemotherapeutic agent mitoxantrone after locoregional cancer treatment. J Drug Target 2010;11:139–49.
[21] Senyei A, Widder K, Czerlinski G. Magnetic guidance of drug-carrying microspheres. J Appl Phys 1978;49(6):3578–83.
[22] Mody VV, Cox A, Shah S, Singh A, Bevins W, Parihar H. Magnetic nanoparticle drug delivery systems for targeting tumor. Appl Nanosci 2014;4(4):385–92.
[23] Arruebo M, Fernández-Pacheco R, Ibarra MR, Santamaría J. Magnetic nanoparticles for drug delivery. Nano today 2007;2(3):22–32.
[24] Dobson J. Magnetic nanoparticles for drug delivery. Drug Develop Res 2006;67 (1):55–60.
[25] Bucak S, Jones DA, Laibinis PE, Hatton TA. Protein separations using colloidal magnetic nanoparticles. Biotechnol Progr 2003;19(2):477–84.
[26] Bucak S, Sharpe S, Kuhn S, Hatton TA. Cell clarification and size separation using continuous countercurrent magnetophoresis. Biotechnol Progr 2011;27(3):744–50.
[27] Lu AH, Salabas EeL, Schüth F. Magnetic nanoparticles: synthesis, protection, functionalization, and application. Angew Chem Int Ed 2007;46(8):1222–44.
[28] Sugimoto T, Matijević E. Formation of uniform spherical magnetite particles by crystallization from ferrous hydroxide gels. J Colloid Interf Sci 1980;74(1):227–43.
[29] Nishio K, Ikeda M, Gokon N, Tsubouchi S, Narimatsu H, Mochizuki Y, et al. Preparation of size-controlled (30–100nm) magnetite nanoparticles for biomedical applications. J Magn Magn Mater 2007;310(2):2408–10.
[30] Vereda F, de Vicente J, Morales MdP, Rull F, Hidalgo-Álvarez R. Synthesis and characterization of single-domain monocrystalline magnetite particles by oxidative aging of $Fe(OH)_2$. J Phys Chem C 2008;112(15):5843–9.
[31] Vereda F, De Vicente J, Hidalgo-Alvarez R. Oxidation of ferrous hydroxides with nitrate: a versatile method for the preparation of magnetic colloidal particles. J Colloid Interf Sci 2013;392:50–6.
[32] Altan CL, Gurten B, Sadza R, Yenigul E, Sommerdijk NA, Bucak S. Poly (acrylic acid)-directed synthesis of colloidally stable single domain magnetite nanoparticles via partial oxidation. J Magn Magn Mater 2016;416:366–72.

[33] Altan CL, Lenders JJ, Bomans PH, de With G, Friedrich H, Bucak S, et al. Partial oxidation as a rational approach to kinetic control in bioinspired magnetite synthesis. Chem A Eur J 2015;21(16):6150−6.

[34] Pradeep T. A textbook of nanoscience and nanotechnology. New Delhi: Tata McGraw-Hill Education; 2012.

[35] Choi H, Choi SR, Zhou R, Kung HF, Chen I-W. Iron oxide nanoparticles as magnetic resonance contrast agent for tumor imaging via folate receptor-targeted delivery 1. Acad Radiol 2004;11(9):996−1004.

[36] Xu C, Xu K, Gu H, Zhong X, Guo Z, Zheng R, et al. Nitrilotriacetic acid-modified magnetic nanoparticles as a general agent to bind histidine-tagged proteins. J Am Chem Soc 2004;126(11):3392−3.

[37] Hadjipanayis CG, Machaidze R, Kaluzova M, Wang L, Schuette AJ, Chen H, et al. EGFRvIII antibody−conjugated iron oxide nanoparticles for magnetic resonance imaging−guided convection-enhanced delivery and targeted therapy of glioblastoma. Cancer Res 2010;70(15):6303−12.

[38] Neuberger T, Schöpf B, Hofmann H, Hofmann M, Von Rechenberg B. Superparamagnetic nanoparticles for biomedical applications: possibilities and limitations of a new drug delivery system. J Magn Magn Mater 2005;293(1):483−96.

[39] Häfeli UO, Riffle JS, Harris-Shekhawat L, Carmichael-Baranauskas A, Mark F, Dailey JP, et al. Cell uptake and in vitro toxicity of magnetic nanoparticles suitable for drug delivery. Mol Pharm 2009;6(5):1417−28.

[40] Yang WJ, Lee JH, Hong SC, Lee J, Lee J, Han D-W. Difference between toxicities of iron oxide magnetic nanoparticles with various surface-functional groups against human normal fibroblasts and fibrosarcoma cells. Materials 2013;6 (10):4689−706.

[41] Kim JS, Yoon T-J, Yu KN, Kim BG, Park SJ, Kim HW, et al. Toxicity and tissue distribution of magnetic nanoparticles in mice. Toxicol Sci 2006;89(1):338−47.

[42] Valdiglesias V, Fernández-Bertólez N, Kiliç G, Costa C, Costa S, Fraga S, et al. Are iron oxide nanoparticles safe? Current knowledge and future perspectives. J Trace Elements Med Biol 2016 (Article in press).

[43] Singh N, Jenkins GJ, Asadi R, Doak SH. Potential toxicity of superparamagnetic iron oxide nanoparticles (SPION). Nano Rev 2010;1.

[44] Park YC, Smith JB, Pham T, Whitaker RD, Sucato CA, Hamilton JA, et al. Effect of PEG molecular weight on stability, T2 contrast, cytotoxicity, and cellular uptake of superparamagnetic iron oxide nanoparticles (SPIONs). Colloids Surf B Biointerfaces 2014;119:106−14.

[45] Zhang Y-N, Poon W, Tavares AJ, McGilvray ID, Chan WC. Nanoparticle−liver interactions: cellular uptake and hepatobiliary elimination. J Control Release 2016 (Article in press).

[46] Sun T, Zhang YS, Pang B, Hyun DC, Yang M, Xia Y. Engineered nanoparticles for drug delivery in cancer therapy. Angew Chem Int Ed 2014;53(46):12320−64.

[47] Yu M, Zheng J. Clearance pathways and tumor targeting of imaging nanoparticles. ACS Nano 2015;9(7):6655−74.

[48] Sun Z, Yathindranath V, Worden M, Thliveris JA, Chu S, Parkinson FE, et al. Characterization of cellular uptake and toxicity of aminosilane-coated iron oxide nanoparticles with different charges in central nervous system-relevant cell culture models. Int J Nanomed 2013;8:961.

[49] Merkel TJ, Jones SW, Herlihy KP, Kersey FR, Shields AR, Napier M, et al. Using mechanobiological mimicry of red blood cells to extend circulation times of hydrogel microparticles. Proc Natl Acad Sci USA 2011;108(2):586–91.
[50] Danhier F, Feron O, Préat V. To exploit the tumor microenvironment: passive and active tumor targeting of nanocarriers for anti-cancer drug delivery. J Control Release 2010;148(2):135–46.
[51] Undevia SD, Gomez-Abuin G, Ratain MJ. Pharmacokinetic variability of anticancer agents. Nat Rev Cancer 2005;5(6):447–58.
[52] Lübbe AS, Bergemann C, Huhnt W, Fricke T, Riess H, Brock JW, et al. Preclinical experiences with magnetic drug targeting: tolerance and efficacy. Cancer Res 1996;56(20):4694–701.
[53] Chari RV. Targeted delivery of chemotherapeutics: tumor-activated prodrug therapy. Adv Drug Deliv Rev 1998;31(1):89–104.
[54] Wahajuddin SA. Superparamagnetic iron oxide nanoparticles: magnetic nanoplatforms as drug carriers. Int J Nanomed 2012;7:3445.
[55] Lim EK, Huh YM, Yang J, Lee K, Suh JS, Haam S. pH-Triggered drug-releasing magnetic nanoparticles for cancer therapy guided by molecular imaging by MRI. Adv Mater 2011;23(21):2436–42.
[56] Chang B, Sha X, Guo J, Jiao Y, Wang C, Yang W. Thermo and pH dual responsive, polymer shell coated, magnetic mesoporous silica nanoparticles for controlled drug release. J Mater Chem 2011;21(25):9239–47.
[57] Gupta P, Vermani K, Garg S. Hydrogels: from controlled release to pH-responsive drug delivery. Drug Discov Today 2002;7(10):569–79.
[58] Yu S, Wu G, Gu X, Wang J, Wang Y, Gao H, et al. Magnetic and pH-sensitive nanoparticles for antitumor drug delivery. Colloids Surf B Biointerfaces 2013;103:15–22.
[59] Zhang W, Gilstrap K, Wu L, Remant Bahadur KC, Moss MA, Wang Q, et al. Synthesis and characterization of thermally responsive pluronic F127 – chitosan nanocapsules for controlled release and intracellular delivery of small molecules. ACS Nano 2010;4(11):6747–59.
[60] Zhang J, Misra R. Magnetic drug-targeting carrier encapsulated with thermosensitive smart polymer: core–shell nanoparticle carrier and drug release response. Acta Biomater 2007;3(6):838–50.
[61] Putnam D, Kopecek J. Enantioselective release of 5-fluorouracil from *N*-(2-hydroxypropyl) methacrylamide-based copolymers via lysosomal enzymes. Bioconjugate Chem 1995;6(4):483–92.
[62] Dubowchik GM, Firestone RA, Padilla L, Willner D, Hofstead SJ, Mosure K, et al. Cathepsin B-labile dipeptide linkers for lysosomal release of doxorubicin from internalizing immunoconjugates: model studies of enzymatic drug release and antigen-specific in vitro anticancer activity. Bioconjugate Chem 2002;13(4):855–69.
[63] Rodzinski A, Guduru R, Liang P, Hadjikhani A, Stewart T, Stimphil E, et al. Targeted and controlled anticancer drug delivery and release with magnetoelectric nanoparticles. Sci Rep 2016;6.
[64] Fan F, Schimming A, Jaeger D, Podar K. Targeting the tumor microenvironment: focus on angiogenesis. J Oncol 2012;2012:281261.
[65] Chertok B, David AE, Yang VC. Brain tumor targeting of magnetic nanoparticles for potential drug delivery: effect of administration route and magnetic field topography. J Control Release 2011;155(3):393–9.

[66] Torchilin VP. Nanoparticulates as drug carriers. London: Imperial College Press; 2006.

[67] Tietze R, Lyer S, Dürr S, Struffert T, Engelhorn T, Schwarz M, et al. Efficient drug-delivery using magnetic nanoparticles—biodistribution and therapeutic effects in tumour bearing rabbits. Nanomed Nanotechnol Biol Med 2013;9(7):961–71.

[68] Bamrungsap S, Zhao Z, Chen T, Wang L, Li C, Fu T, et al. Nanotechnology in therapeutics: a focus on nanoparticles as a drug delivery system. Nanomedicine 2012;7(8):1253–71.

[69] Singh A, Sahoo SK. Magnetic nanoparticles: a novel platform for cancer theranostics. Drug Discov Today 2014;19(4):474–81.

[70] Hola K, Markova Z, Zoppellaro G, Tucek J, Zboril R. Tailored functionalization of iron oxide nanoparticles for MRI, drug delivery, magnetic separation and immobilization of biosubstances. Biotechnol Adv 2015;33(6):1162–76.

[71] Alexiou C, Arnold W, Klein RJ, Parak FG, Hulin P, Bergemann C, et al. Locoregional cancer treatment with magnetic drug targeting. Cancer Res 2000;60(23):6641–8.

[72] Lübbe AS, Bergemann C, Riess H, Schriever F, Reichardt P, Possinger K, et al. Clinical experiences with magnetic drug targeting: a phase I study with 4′-epidoxorubicin in 14 patients with advanced solid tumors. Cancer Res 1996;56(20):4686–93.

[73] Alexiou C, Schmid RJ, Jurgons R, Kremer M, Wanner G, Bergemann C, et al. Targeting cancer cells: magnetic nanoparticles as drug carriers. Eur Biophys J 2006;35(5):446–50.

[74] Chertok B, Moffat BA, David AE, Yu F, Bergemann C, Ross BD, et al. Iron oxide nanoparticles as a drug delivery vehicle for MRI monitored magnetic targeting of brain tumors. Biomaterials 2008;29(4):487–96.

[75] Parveen S, Misra R, Sahoo SK. Nanoparticles: a boon to drug delivery, therapeutics, diagnostics and imaging. Nanomed Nanotechnol Biol Med 2012;8(2):147–66.

[76] Voltairas P, Fotiadis D, Michalis L. Hydrodynamics of magnetic drug targeting. J Biomech 2002;35(6):813–21.

[77] Zhang Y, Kohler N, Zhang M. Surface modification of superparamagnetic magnetite nanoparticles and their intracellular uptake. Biomaterials 2002;23(7):1553–61.

[78] Kayal S, Ramanujan R. Doxorubicin loaded PVA coated iron oxide nanoparticles for targeted drug delivery. Mater Sci Eng C 2010;30(3):484–90.

[79] Kohler N, Sun C, Fichtenholtz A, Gunn J, Fang C, Zhang M. Methotrexate-immobilized poly(ethylene glycol) magnetic nanoparticles for MR imaging and drug delivery. Small 2006;2(6):785–92.

[80] Yallapu MM, Foy SP, Jain TK, Labhasetwar V. PEG-functionalized magnetic nanoparticles for drug delivery and magnetic resonance imaging applications. Pharm Res 2010;27(11):2283–95.

[81] Yuan Q, Venkatasubramanian R, Hein S, Misra R. A stimulus-responsive magnetic nanoparticle drug carrier: magnetite encapsulated by chitosan-grafted-copolymer. Acta Biomater 2008;4(4):1024–37.

[82] Hassan EE, Parish RC, Gallo JM. Optimized formulation of magnetic chitosan microspheres containing the anticancer agent, oxantrazole. Pharm Res 1992;9(3):390–7.

[83] Chertok B, David AE, Yang VC. Polyethyleneimine-modified iron oxide nanoparticles for brain tumor drug delivery using magnetic targeting and intra-carotid administration. Biomaterials 2010;31(24):6317–24.

[84] Alexiou C, Jurgons R, Schmid R, Hilpert A, Bergemann C, Parak F, et al. In vitro and in vivo investigations of targeted chemotherapy with magnetic nanoparticles. J Magn Magn Mater 2005;293(1):389–93.

[85] Shinkai M, Suzuki M, Iijima S, Kobayashi T. Antibody-conjugated magnetoliposomes for targeting cancer cells and their application in hyperthermia. Biotechnol Appl Biochem 1995;21(2):125–37.

[86] Mohapatra S, Mallick S, Maiti T, Ghosh S, Pramanik P. Synthesis of highly stable folic acid conjugated magnetite nanoparticles for targeting cancer cells. Nanotechnology 2007;18(38):385102.

[87] Wang AZ, Bagalkot V, Vasilliou CC, Gu F, Alexis F, Zhang L, et al. Superparamagnetic iron oxide nanoparticle–aptamer bioconjugates for combined prostate cancer imaging and therapy. ChemMedChem 2008;3(9):1311–15.

[88] Torchilin VP. Recent advances with liposomes as pharmaceutical carriers. Nat Rev Drug Discov 2005;4(2):145–60.

[89] Zhang J, Zhang Z, Yang H, Tan Q, Qin S, Qiu X. Lyophilized paclitaxel magnetoliposomes as a potential drug delivery system for breast carcinoma via parenteral administration: in vitro and in vivo studies. Pharm Res 2005;22(4):573–83.

[90] Chen F-H, Zhang L-M, Chen Q-T, Zhang Y, Zhang Z-J. Synthesis of a novel magnetic drug delivery system composed of doxorubicin-conjugated Fe_3O_4 nanoparticle cores and a PEG-functionalized porous silica shell. Chem Commun 2010;46 (45):8633–5.

[91] Wu H, Liu G, Zhang S, Shi J, Zhang L, Chen Y, et al. Biocompatibility, MR imaging and targeted drug delivery of a rattle-type magnetic mesoporous silica nanosphere system conjugated with PEG and cancer-cell-specific ligands. J Mater Chem 2011;21 (9):3037–45.

[92] Kim J, Kim HS, Lee N, Kim T, Kim H, Yu T, et al. Multifunctional uniform nanoparticles composed of a magnetite nanocrystal core and a mesoporous silica shell for magnetic resonance and fluorescence imaging and for drug delivery. Angew Chem Int Ed 2008;47(44):8438–41.

[93] Yuan Y, Rende D, Altan CL, Bucak S, Ozisik R, Borca-Tasciuc D-A. Effect of surface modification on magnetization of iron oxide nanoparticle colloids. Langmuir 2012;28(36):13051–9.

[94] Mosbach K, Schröder U. Preparation and application of magnetic polymers for targeting of drugs. FEBS Lett 1979;102(1):112–16.

[95] Singh A, Dilnawaz F, Mewar S, Sharma U, Jagannathan N, Sahoo SK. Composite polymeric magnetic nanoparticles for co-delivery of hydrophobic and hydrophilic anticancer drugs and MRI imaging for cancer therapy. ACS Appl Mater Interf 2011;3(3):842–56.

[96] Shen J-M, Gao F-Y, Yin T, Zhang H-X, Ma M, Yang Y-J, et al. cRGD-functionalized polymeric magnetic nanoparticles as a dual-drug delivery system for safe targeted cancer therapy. Pharmacol Res 2013;70(1):102–15.

[97] Sun C, Fang C, Stephen Z, Veiseh O, Hansen S, Lee D, et al. Tumor-targeted drug delivery and MRI contrast enhancement by chlorotoxin-conjugated iron oxide nanoparticles. Nanomedicine (Lond) 2008;3(4):495–505.

[98] Ma X, Tao H, Yang K, Feng L, Cheng L, Shi X, et al. A functionalized graphene oxide-iron oxide nanocomposite for magnetically targeted drug delivery, photothermal therapy, and magnetic resonance imaging. Nano Res 2012;5(3):199–212.

[99] Mamelak AN, Jacoby DB. Targeted delivery of antitumoral therapy to glioma and other malignancies with synthetic chlorotoxin (TM-601). Expert Opin Drug Deliv 2007;4(2):175–86.

[100] Grief AD, Richardson G. Mathematical modelling of magnetically targeted drug delivery. J Magn Magn Mater 2005;293(1):455–63.

[101] Christophi C, Winkworth A, Muralihdaran V, Evans P. The treatment of malignancy by hyperthermia. Surg Oncol 1998;7(1-2):83–90.

[102] Harmon BV, Takano YS, Winterford CM, Gobe GC. The role of apoptosis in the response of cells and tumours to mild hyperthermia. Int J Radiat Biol 1991;59 (2):489–501.

[103] Sellins KS, Cohen JJ. Hyperthermia induces apoptosis in thymocytes. Radiat Res 1991;126(1):88–95.

[104] Moroz P, Jones SK, Gray BN. Status of hyperthermia in the treatment of advanced liver cancer. J Surg Oncol 2001;77(4):259–69.

[105] Kumar CS, Mohammad F. Magnetic nanomaterials for hyperthermia-based therapy and controlled drug delivery. Adv Drug Deliv Rev 2011;63(9):789–808.

[106] Habash RW, Bansal R, Krewski D, Alhafid HT. Thermal therapy, part 2: hyperthermia techniques. Crit Rev Biomed Eng 2006;34(6).

[107] Thiesen B, Jordan A. Clinical applications of magnetic nanoparticles for hyperthermia. Int J Hyperthermia 2008;24(6):467–74.

[108] Bornstein BA, Zouranjian PS, Hansen JL, Fraser SM, Gelwan LA, Teicher BA, et al. Local hyperthermia, radiation therapy, and chemotherapy in patients with local-regional recurrence of breast carcinoma. Int J Radiat Oncol Biol Phys 1993;25 (1):79–85.

[109] Group ICH, Vernon CC, Hand JW, Field SB, Machin D, Whaley JB, et al. Radiotherapy with or without hyperthermia in the treatment of superficial localized breast cancer: results from five randomized controlled trials. Int J Radiat Oncol Biol Phys 1996;35(4):731–44.

[110] Overgaard J, Bentzen S, Gonzalez DG, Hulshof M, Arcangeli G, Dahl O, et al. Randomised trial of hyperthermia as adjuvant to radiotherapy for recurrent or metastatic malignant melanoma. Lancet 1995;345(8949):540–3.

[111] van der Zee J, González D, van Rhoon GC, van Dijk JD, van Putten WL, Hart AA. Comparison of radiotherapy alone with radiotherapy plus hyperthermia in locally advanced pelvic tumours: a prospective, randomised, multicentre trial. Lancet 2000;355(9210):1119–25.

[112] Wust P, Hildebrandt B, Sreenivasa G, Rau B, Gellermann J, Riess H, et al. Hyperthermia in combined treatment of cancer. Lancet Oncol 2002;3(8):487–97.

[113] Deatsch AE, Evans BA. Heating efficiency in magnetic nanoparticle hyperthermia. J Magn Magn Mater 2014;354:163–72.

[114] Hergt R, Andra W, d'Ambly CG, Hilger I, Kaiser WA, Richter U, et al. Physical limits of hyperthermia using magnetite fine particles. IEEE Trans Magn 1998;34 (5):3745–54.

[115] Gonzales-Weimuller M, Zeisberger M, Krishnan KM. Size-dependant heating rates of iron oxide nanoparticles for magnetic fluid hyperthermia. J Magn Magn Mater 2009;321(13):1947–50.

[116] Gilchrist R, Medal R, Shorey WD, Hanselman RC, Parrott JC, Taylor CB. Selective inductive heating of lymph nodes. Ann Surg 1957;146(4):596.

[117] Gordon R, Hines J, Gordon D. Intracellular hyperthermia. A biophysical approach to cancer treatment via intracellular temperature and biophysical alterations. Med Hypotheses 1979;5(1):83–102.
[118] Chan DC, Kirpotin DB, Bunn PA. Synthesis and evaluation of colloidal magnetic iron oxides for the site-specific radiofrequency-induced hyperthermia of cancer. J Magn Magn Mater 1993;122(1):374–8.
[119] Jordan A, Scholz R, Wust P, Fähling H, Krause J, Wlodarczyk W, et al. Effects of magnetic fluid hyperthermia (MFH) on C3H mammary carcinoma in vivo. Int J Hyperthermia 1997;13(6):587–605.
[120] Mornet S, Vasseur S, Grasset F, Duguet E. Magnetic nanoparticle design for medical diagnosis and therapy. J Mater Chem 2004;14(14):2161–75.
[121] Barry SE. Challenges in the development of magnetic particles for therapeutic applications. Int J Hyperthermia 2008;24(6):451–66.
[122] Nagesha D, Devalapally H, Sridhar S, Amiji MM. Multifunctional magnetic nanosystems for tumor imaging, targeted delivery, and thermal medicine. Multifunctional Pharmaceutical Nanocarriers. New York, NY: Springer; 2008. p. 381–408.
[123] Zhang X, Chen S, Wang H-M, Hsieh S-L, Wu C-H, Chou H-H, et al. Role of Néel and Brownian relaxation mechanisms for water-based Fe_3O_4 nanoparticle ferrofluids in hyperthermia. Biomed Eng Appl Basis Commun 2010;22(05):393–9.
[124] Wildeboer RR, Southern P, Pankhurst QA. On the reliable measurement of specific absorption rates and intrinsic loss parameters in magnetic hyperthermia materials. J Phys D Appl Phys 2014;47(49):495003.
[125] Kim DH, Kim KN, Kim KM, Lee YK. Targeting to carcinoma cells with chitosan- and starch-coated magnetic nanoparticles for magnetic hyperthermia. J Biomed Mater Res A 2009;88(1):1–11.
[126] Kim M-J, Jang D-H, Lee Y-I, Jung HS, Lee H-J, Choa Y-H. Preparation, characterization, cytotoxicity and drug release behavior of liposome-enveloped paclitaxel/Fe_3O_4 nanoparticles. J Nanosci Nanotechnol 2011;11(1):889–93.
[127] Hua M-Y, Yang H-W, Chuang C-K, Tsai R-Y, Chen W-J, Chuang K-L, et al. Magnetic-nanoparticle-modified paclitaxel for targeted therapy for prostate cancer. Biomaterials 2010;31(28):7355–63.
[128] Hergt R, Hiergeist R, Zeisberger M, Schüler D, Heyen U, Hilger I, et al. Magnetic properties of bacterial magnetosomes as potential diagnostic and therapeutic tools. J Magn Magn Mater 2005;293(1):80–6.
[129] Hergt R, Hiergeist R, Zeisberger M, Glöckl G, Weitschies W, Ramirez L, et al. Enhancement of AC-losses of magnetic nanoparticles for heating applications. J Magn Magn Mater 2004;280(2):358–68.
[130] Hayashi K, Nakamura M, Sakamoto W, Yogo T, Miki H, Ozaki S, et al. Superparamagnetic nanoparticle clusters for cancer theranostics combining magnetic resonance imaging and hyperthermia treatment. Theranostics 2013;3(6):366–76.
[131] Asin L, Ibarra M, Tres A, Goya G. Controlled cell death by magnetic hyperthermia: effects of exposure time, field amplitude, and nanoparticle concentration. Pharm Res 2012;29(5):1319–27.
[132] Jordan A, Scholz R, Wust P, Schirra H, Schiestel T, Schmidt H, et al. Endocytosis of dextran and silan-coated magnetite nanoparticles and the effect of intracellular hyperthermia on human mammary carcinoma cells in vitro. J Magn Magn Mater 1999;194(1):185–96.

[133] Jadhav NV, Prasad AI, Kumar A, Mishra R, Dhara S, Babu K, et al. Synthesis of oleic acid functionalized Fe_3O_4 magnetic nanoparticles and studying their interaction with tumor cells for potential hyperthermia applications. Colloids Surf B Biointerfaces 2013;108:158–68.

[134] Guardia P, Di Corato R, Lartigue L, Wilhelm C, Espinosa A, Garcia-Hernandez M, et al. Water-soluble iron oxide nanocubes with high values of specific absorption rate for cancer cell hyperthermia treatment. ACS Nano 2012;6(4):3080–91.

[135] Hayashi K, Ono K, Suzuki H, Sawada M, Moriya M, Sakamoto W, et al. One-pot biofunctionalization of magnetic nanoparticles via thiol–ene click reaction for magnetic hyperthermia and magnetic resonance imaging. Chem Mater 2010;22(12):3768–72.

[136] Ghosh R, Pradhan L, Devi YP, Meena S, Tewari R, Kumar A, et al. Induction heating studies of Fe_3O_4 magnetic nanoparticles capped with oleic acid and polyethylene glycol for hyperthermia. J Mater Chem 2011;21(35):13388–98.

[137] Samanta B, Yan H, Fischer NO, Shi J, Jerry DJ, Rotello VM. Protein-passivated Fe_3O_4 nanoparticles: low toxicity and rapid heating for thermal therapy. J Mater Chem 2008;18(11):1204–8.

[138] Maier-Hauff K, Rothe R, Scholz R, Gneveckow U, Wust P, Thiesen B, et al. Intracranial thermotherapy using magnetic nanoparticles combined with external beam radiotherapy: results of a feasibility study on patients with glioblastoma multiforme. J Neuro-Oncol 2007;81(1):53–60.

[139] Dennis C, Jackson A, Borchers J, Ivkov R, Foreman A, Hoopes P, et al. The influence of magnetic and physiological behaviour on the effectiveness of iron oxide nanoparticles for hyperthermia. J Phys D Appl Phys 2008;41(13):134020.

[140] Pradhan P, Giri J, Banerjee R, Bellare J, Bahadur D. Cellular interactions of lauric acid and dextran-coated magnetite nanoparticles. J Magn Magn Mater 2007;311(1):282–7.

[141] Kawai N, Futakuchi M, Yoshida T, Ito A, Sato S, Naiki T, et al. Effect of heat therapy using magnetic nanoparticles conjugated with cationic liposomes on prostate tumor in bone. Prostate 2008;68(7):784–92.

[142] de Sousa ME, Fernández van Raap MB, Rivas PC, Mendoza Zélis P, Girardin P, Pasquevich GA, et al. Stability and relaxation mechanisms of citric acid coated magnetite nanoparticles for magnetic hyperthermia. J Phys Chem C 2013;117(10):5436–45.

[143] Martín-Saavedra F, Ruíz-Hernández E, Boré A, Arcos D, Vallet-Regí M, Vilaboa N. Magnetic mesoporous silica spheres for hyperthermia therapy. Acta Biomater 2010;6(12):4522–31.

[144] Huang HS, Hainfeld JF. Intravenous magnetic nanoparticle cancer hyperthermia. Int J Nanomed 2013;8:2521.

[145] Gutiérrez L, Mejías R, Barber D, Veintemillas-Verdaguer S, Serna C, Lázaro F, et al. Ac magnetic susceptibility study of in vivo nanoparticle biodistribution. J Phys D Appl Phys 2011;44(25):255002.

[146] Yanase M, Shinkai M, Honda H, Wakabayashi T, Yoshida J, Kobayashi T. Antitumor immunity induction by intracellular hyperthermia using magnetite cationic liposomes. Jpn J Cancer Res 1998;89(7):775–82.

[147] Matsuoka F, Shinkai M, Honda H, Kubo T, Sugita T, Kobayashi T. Hyperthermia using magnetite cationic liposomes for hamster osteosarcoma. Biomagn Res Technol 2004;2:3.

[148] Lee J-H, Jang J-T, Choi J-S, Moon SH, Noh S-H, Kim J-W, et al. Exchange-coupled magnetic nanoparticles for efficient heat induction. Nat Nanotechnol 2011;6(7):418–22.
[149] Balivada S, Rachakatla RS, Wang H, Samarakoon TN, Dani RK, Pyle M, et al. A/C magnetic hyperthermia of melanoma mediated by iron (0)/iron oxide core/shell magnetic nanoparticles: a mouse study. BMC Cancer 2010;10(1):1.
[150] Hu H, Sheng Y, Ye M, Qian Y, Tang J, Shen Y. A porphyrin-based magnetic and fluorescent dual-modal nanoprobe for tumor imaging. Polymer 2016;88:94–101.
[151] Cheung ENM, Alvares RD, Oakden W, Chaudhary R, Hill ML, Pichaandi J, et al. Polymer-stabilized lanthanide fluoride nanoparticle aggregates as contrast agents for magnetic resonance imaging and computed tomography. Chem Mater 2010;22(16):4728–39.
[152] Ghaghada KB, Ravoori M, Sabapathy D, Bankson J, Kundra V, Annapragada A. New dual mode gadolinium nanoparticle contrast agent for magnetic resonance imaging. PLoS ONE 2009;4(10):e7628.
[153] Lu J, Ma S, Sun J, Xia C, Liu C, Wang Z, et al. Manganese ferrite nanoparticle micellar nanocomposites as MRI contrast agent for liver imaging. Biomaterials 2009;30(15):2919–28.
[154] Choi JS, Park JC, Nah H, Woo S, Oh J, Kim KM, et al. A hybrid nanoparticle probe for dual-modality positron emission tomography and magnetic resonance imaging. Angew Chem Int Ed 2008;47(33):6259–62.
[155] Sattarahmady N, Zare T, Mehdizadeh A, Azarpira N, Heidari M, Lotfi M, et al. Dextrin-coated zinc substituted cobalt-ferrite nanoparticles as an MRI contrast agent: In vitro and in vivo imaging studies. Colloids Surf B Biointerfaces 2015;129:15–20.
[156] Choi J-s, Choi HJ, Jung DC, Lee J-H, Cheon J. Nanoparticle assisted magnetic resonance imaging of the early reversible stages of amyloid β self-assembly. Chem Commun 2008;19:2197–9.
[157] An L, Yu Y, Li X, Liu W, Yang H, Wu D, et al. Dextran-coated superparamagnetic amorphous Fe–Co nanoalloy for magnetic resonance imaging applications. Mater Res Bull 2014;49:285–90.
[158] Xu YH, Bai J, Wang J-P. High-magnetic-moment multifunctional nanoparticles for nanomedicine applications. J Magn Magn Mater 2007;311(1):131–4.
[159] Chung R-J, Ou K-L, Chen S-P, Liu H-L. Preparation of ICG-FePt nanoparticles promising for magnetic resonance imaging contrast agent and hyperthermia applications. Adv Powder Technol 2016;27:994–9.
[160] Maenosono S, Suzuki T, Saita S. Superparamagnetic FePt nanoparticles as excellent MRI contrast agents. J Magn Magn Mater 2008;320(9):L79–83.
[161] Schinhammer M, Gerber I, Hänzi AC, Uggowitzer PJ. On the cytocompatibility of biodegradable Fe-based alloys. Mater Sci Eng C 2013;33(2):782–9.
[162] Bagchi D, Stohs SJ, Downs BW, Bagchi M, Preuss HG. Cytotoxicity and oxidative mechanisms of different forms of chromium. Toxicology 2002;180(1):5–22.
[163] Charlton JR, Pearl VM, Denotti AR, Lee JB, Swaminathan S, Scindia YM, et al. Biocompatibility of ferritin-based nanoparticles as targeted MRI contrast agents. Nanomed Nanotechnol Biol Med 2016;12(6):1735–45.
[164] Kılınç E. γ-Fe_2O_3 magnetic nanoparticle functionalized with carboxylated multi walled carbon nanotube: synthesis, characterization, analytical and biomedical application. J Magn Magn Mater 2016;401:949–55.

[165] Chen Y, Gu H. Microwave assisted fast fabrication of Fe_3O_4-MWCNTs nanocomposites and their application as MRI contrast agents. Mater Lett 2012;67(1):49–51.

[166] Liu Y, Hughes TC, Muir BW, Waddington LJ, Gengenbach TR, Easton CD, et al. Water-dispersible magnetic carbon nanotubes as T2-weighted MRI contrast agents. Biomaterials 2014;35(1):378–86.

[167] Ding W, Lou C, Qiu J, Zhao Z, Zhou Q, Liang M, et al. Targeted Fe-filled carbon nanotube as a multifunctional contrast agent for thermoacoustic and magnetic resonance imaging of tumor in living mice. Nanomed Nanotechnol Biol Med 2016;12 (1):235–44.

[168] Saokar A, Gee MS, Islam T, Mueller PR, Harisinghani MG. Appearance of primary lymphoid malignancies on lymphotropic nanoparticle-enhanced magnetic resonance imaging using ferumoxtran-10. Clin Imaging 2010;34(6):448–52.

[169] Guimaraes AR, Tabatabei S, Dahl D, McDougal WS, Weissleder R, Harisinghani MG. Pilot study evaluating use of lymphotrophic nanoparticle-enhanced magnetic resonance imaging for assessing lymph nodes in renal cell cancer. Urology 2008;71(4):708–12.

[170] Wang Y-XJ. Superparamagnetic iron oxide based MRI contrast agents: current status of clinical application. Quant Imaging Med Surg 2011;1(1):35–40.

[171] Toso C, Vallee JP, Morel P, Ris F, Demuylder-Mischler S, Lepetit-Coiffe M, et al. Clinical magnetic resonance imaging of pancreatic islet grafts after iron nanoparticle labeling. Am J Transplant 2008;8(3):701–6.

[172] Yilmaz A, Rösch S, Klingel K, Kandolf R, Helluy X, Hiller K-H, et al. Magnetic resonance imaging (MRI) of inflamed myocardium using iron oxide nanoparticles in patients with acute myocardial infarction—preliminary results. Int J Cardiol 2013;163(2):175–82.

[173] Chao Y, Makale M, Karmali PP, Sharikov Y, Tsigelny I, Merkulov S, et al. Recognition of dextran–superparamagnetic iron oxide nanoparticle conjugates (Feridex) via macrophage scavenger receptor charged domains. Bioconjugate Chem 2012;23(5):1003–9.

[174] Vilarino-Varela MJ, Taylor A, Rockall AG, Reznek RH, Powell ME. A verification study of proposed pelvic lymph node localisation guidelines using nanoparticle-enhanced magnetic resonance imaging. Radiother Oncol 2008;89(2):192–6.

[175] Klein C, Nagel E, Schnackenburg B, Bornstedt A, Schalla S, Hoffmann V, et al. The intravascular contrast agent Clariscan™ (NC 100150 injection) for 3D MR coronary angiography in patients with coronary artery disease. Magn Reson Mater Phys Biol Med 2000;11(1–2):65–7.

[176] Jung CW, Jacobs P. Physical and chemical properties of superparamagnetic iron oxide MR contrast agents: ferumoxides, ferumoxtran, ferumoxsil. Magn Reson Imaging 1995;13(5):661–74.

[177] Achiam M, Løgager V, Chabanova E, Thomsen H, Rosenberg J. Patient acceptance of MR colonography with improved fecal tagging versus conventional colonoscopy. Eur J Radiol 2010;73(1):143–7.

[178] Elsherbini AA, Saber M, Aggag M, El-Shahawy A, Shokier HA. Magnetic nanoparticle-induced hyperthermia treatment under magnetic resonance imaging. Magn Reson Imaging 2011;29(2):272–80.

[179] Soares PI, Laia CA, Carvalho A, Pereira LC, Coutinho JT, Ferreira IM, et al. Iron oxide nanoparticles stabilized with a bilayer of oleic acid for magnetic hyperthermia and MRI applications. Appl Surf Sci 2016;383:240–7.

[180] Guldris N, Argibay B, Kolen'ko YV, Carbó-Argibay E, Sobrino T, Campos F, et al. Influence of the separation procedure on the properties of magnetic nanoparticles: Gaining in vitro stability and T 1–T 2 magnetic resonance imaging performance. J Colloid Interface Sci 2016;472:229–36.

[181] Wilhelm C, Gazeau F. Universal cell labelling with anionic magnetic nanoparticles. Biomaterials 2008;29(22):3161–74.

[182] Villanueva A, Canete M, Roca AG, Calero M, Veintemillas-Verdaguer S, Serna CJ, et al. The influence of surface functionalization on the enhanced internalization of magnetic nanoparticles in cancer cells. Nanotechnology 2009;20(11):115103.

[183] Uchiyama MK, Toma SH, de Paula Rodrigues SF, Shimada ALB, Loiola RA, Rodríguez HJC, et al. Ultrasmall cationic superparamagnetic iron oxide nanoparticles as nontoxic and efficient MRI contrast agent and magnetic-targeting tool. Int J Nanomed 2015;10:4731.

[184] Zhao Z, Zhou Z, Bao J, Wang Z, Hu J, Chi X, et al. Octapod iron oxide nanoparticles as high-performance T2 contrast agents for magnetic resonance imaging. Nat Commun 2013;4:2266.

[185] Ruiz A, Salas G, Calero M, Hernández Y, Villanueva A, Herranz F, et al. Short-chain PEG molecules strongly bound to magnetic nanoparticle for MRI long circulating agents. Acta Biomater 2013;9(5):6421–30.

[186] Trekker J, Leten C, Struys T, Lazenka VV, Argibay B, Micholt L, et al. Sensitive in vivo cell detection using size-optimized superparamagnetic nanoparticles. Biomaterials 2014;35(5):1627–35.

[187] Kokuryo D, Anraku Y, Kishimura A, Tanaka S, Kano MR, Kershaw J, et al. SPIO-PICsome: Development of a highly sensitive and stealth-capable MRI nano-agent for tumor detection using SPIO-loaded unilamellar polyion complex vesicles (PICsomes). J Control Release 2013;169(3):220–7.

[188] Semkina A, Abakumov M, Grinenko N, Abakumov A, Skorikov A, Mironova E, et al. Core–shell–corona doxorubicin-loaded superparamagnetic Fe_3O_4 nanoparticles for cancer theranostics. Colloids Surf B Biointerfaces 2015;136:1073–80.

[189] Han L, Zhou X. Synthesis and characterization of liposomes nano-composite-particles with hydrophobic magnetite as a MRI probe. Appl Surf Sci 2016;376:252–60.

[190] Nguyen DH, Lee JS, Choi JH, Park KM, Lee Y, Park KD. Hierarchical self-assembly of magnetic nanoclusters for theranostics: tunable size, enhanced magnetic resonance imagability, and controlled and targeted drug delivery. Acta Biomater 2016;35:109–17.

[191] Rahn H, Woodward R, House M, Engineer D, Feindel K, Dutz S, et al. Calibration standard of body tissue with magnetic nanocomposites for MRI and X-ray imaging. J Magn Magn Mater 2016;405:78–87.

[192] Zhou C-j, Wang S-h, Yu Z, Rong P-f, Chen Z-z, Liu J-y, et al. Folate-conjugated Fe_3O_4 nanoparticles for in vivo tumor labeling. Trans Nonferrous Met Soc China 2013;23(7):2079–84.

[193] Sanjai C, Kothan S, Gonil P, Saesoo S, Sajomsang W. Super-paramagnetic loaded nanoparticles based on biological macromolecules for in vivo targeted MR imaging. Int J Biol Macromol 2016;86:233–41.

[194] Sanjai C, Kothan S, Gonil P, Saesoo S, Sajomsang W. Chitosan-triphosphate nanoparticles for encapsulation of super-paramagnetic iron oxide as an MRI contrast agent. Carbohydr Polym 2014;104:231–7.

[195] Jiang Q, Zheng S, Hong R, Deng S, Guo L, Hu R, et al. Folic acid-conjugated Fe_3O_4 magnetic nanoparticles for hyperthermia and MRI in vitro and in vivo. Appl Surf Sci 2014;307:224–33.

[196] Gu J, Zhang W, Yang X. Preparation of a superparamagnetic MRI contrast agent with a tumor targeting function. Mater Lett 2013;94:8–10.

[197] Lee J, Kim KS, Na K. Caffeic acid-coated multifunctional magnetic nanoparticles for the treatment and bimodal imaging of tumours. J Photochem Photobiol B Biol 2016;160:210–16.

[198] Ma Q, Nakane Y, Mori Y, Hasegawa M, Yoshioka Y, Watanabe TM, et al. Multilayered, core/shell nanoprobes based on magnetic ferric oxide particles and quantum dots for multimodality imaging of breast cancer tumors. Biomaterials 2012;33(33):8486–94.

[199] Yazdani F, Fattahi B, Azizi N. Synthesis of functionalized magnetite nanoparticles to use as liver targeting MRI contrast agent. J Magn Magn Mater 2016;406:207–11.

[200] Rosenberger I, Strauss A, Dobiasch S, Weis C, Szanyi S, Gil-Iceta L, et al. Targeted diagnostic magnetic nanoparticles for medical imaging of pancreatic cancer. J Control Release 2015;214:76–84.

[201] Meng X-x Wan J-q, Jing M, Zhao S-g, Cai W, Liu E-z. Specific targeting of gliomas with multifunctional superparamagnetic iron oxide nanoparticle optical and magnetic resonance imaging contrast agents. Acta Pharmacol Sin 2007;28(12):2019.

[202] Azhdarzadeh M, Atyabi F, Saei AA, Varnamkhasti BS, Omidi Y, Fateh M, et al. Theranostic MUC-1 aptamer targeted gold coated superparamagnetic iron oxide nanoparticles for magnetic resonance imaging and photothermal therapy of colon cancer. Colloids Surf B Biointerfaces 2016;143:224–32.

[203] Jaganathan H, Ivanisevic A. Gold–iron oxide nanoparticle chains scaffolded on DNA as potential magnetic resonance imaging agents. J Mater Chem 2011;21(4):939–43.

[204] Qin J, Li K, Peng C, Li X, Lin J, Ye K, et al. MRI of iron oxide nanoparticle-labeled ADSCs in a model of hindlimb ischemia. Biomaterials 2013;34(21):4914–25.

[205] Egawa EY, Kitamura N, Nakai R, Arima Y, Iwata H. A DNA hybridization system for labeling of neural stem cells with SPIO nanoparticles for MRI monitoring post-transplantation. Biomaterials 2015;54:158–67.

[206] Wang C, Ravi S, Martinez GV, Chinnasamy V, Raulji P, Howell M, et al. Dual-purpose magnetic micelles for MRI and gene delivery. J Control Release 2012;163(1):82–92.

[207] Cheng Z, Dai Y, Kang X, Li C, Huang S, Lian H, et al. Gelatin-encapsulated iron oxide nanoparticles for platinum (IV) prodrug delivery, enzyme-stimulated release and MRI. Biomaterials 2014;35(24):6359–68.

[208] Huang J, Li Y, Orza A, Lu Q, Guo P, Wang L, et al. Magnetic nanoparticle facilitated drug delivery for cancer therapy with targeted and image-guided approaches. Adv Funct Mater 2016;26:3818–36.

[209] Xu H, Cheng L, Wang C, Ma X, Li Y, Liu Z. Polymer encapsulated upconversion nanoparticle/iron oxide nanocomposites for multimodal imaging and magnetic targeted drug delivery. Biomaterials 2011;32(35):9364–73.

[210] Yallapu MM, Othman SF, Curtis ET, Gupta BK, Jaggi M, Chauhan SC. Multi-functional magnetic nanoparticles for magnetic resonance imaging and cancer therapy. Biomaterials 2011;32(7):1890–905.

[211] Thomas RG, Moon MJ, Lee H, Sasikala ARK, Kim CS, Park I-K, et al. Hyaluronic acid conjugated superparamagnetic iron oxide nanoparticle for cancer diagnosis and hyperthermia therapy. Carbohydr Polym 2015;131:439–46.

CHAPTER

Nanotechnology in neurosurgical oncology

7

A. Tan[1,2,3,*], R. Jeyaraj[1,*] and S.F. De Lacey[3,4]

[1]University College London (UCL), London, United Kingdom [2]Stanford University, Stanford, CA, United States [3]Kids Development Foundation, London, United Kingdom [4]University of Zurich, Zurich, Switzerland

CHAPTER OUTLINE

*Equal first authors

BACKGROUND

INTRODUCTION TO BRAIN TUMORS

While the incidence rates of most cancers have decreased or remained stable over the past several decades [1], the age-adjusted incidence rate of brain tumors has increased [2,3]. It is unclear if this increase is due to a true rise in disease incidence, or if it is instead a natural result of improved diagnosis. What is clear, however, is that brain tumors remain notoriously difficult to treat. In fact, they are the leading cause of cancer-related death in patients younger than 35 years of age, and mortality rates increase progressively with age (Fig. 7.1) [4].

Brain tumors can be broadly classified into primary or secondary tumors. Primary tumors represent over 100 different tumor types with unique biologies [5]. Half of these arise from cells within the brain itself and are known as intrinsic lesions, while the remainder originate in the meninges or nerves and are known as extrinsic lesions. The majority of primary brain tumors are intrinsic lesions that arise from supporting glial cells; they are broadly referred to as gliomas (Fig. 7.2) [6].

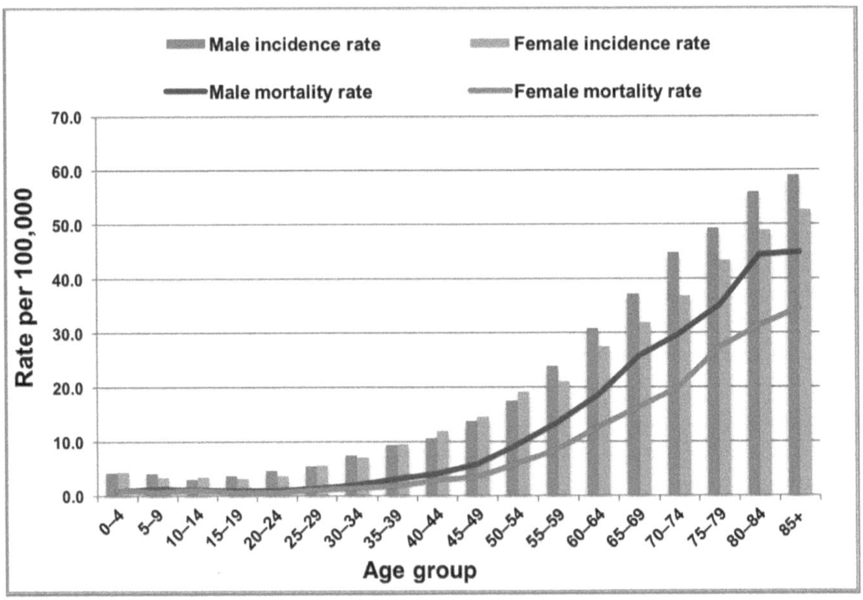

FIGURE 7.1

Age-specific incidence and mortality rates of brain and other CNS neoplasms per 100,000 population in the United Kingdom.

Adapted from Cancer Research UK, http://www.cancerresearchuk.org/health-professional/cancer-statistics/statistics-by-cancer-type/brain-other-cns-and-intracranial-tumours/incidence#heading-One. Accessed April 11, 2016. Copyright © 2016 Cancer Research UK. Reproduced with permission from Cancer Research UK.

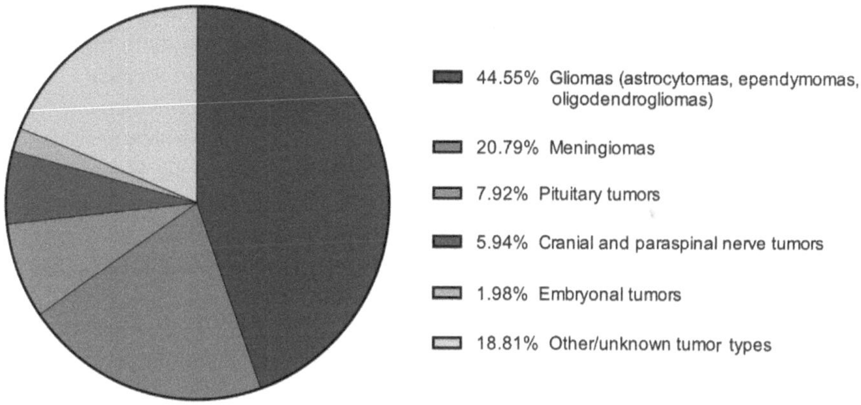

FIGURE 7.2

Proportion of cases of brain, other CNS, and intracranial tumors by morphology, England, 2006–10.

Adapted from Cancer Research UK, http://www.cancerresearchuk.org/health-professional/cancer-statistics/statistics-by-cancer-type/brain-other-cns-and-intracranial-tumours/incidence#heading-One. Accessed April 11, 2016. Copyright © 2016 Cancer Research UK. Reproduced with permission from Cancer Research UK.

Secondary brain tumors arise from cancer cells that have metastasized from a different area of the body—often from the lung, breast, colon, kidney, and skin—and deposited in the brain. These secondary brain metastases occur in 20–40% of all patients with cancer.

Despite their diverse etiologies, all brain tumors are relatively resistant to currently available treatment options. The most common primary brain cancer, glioblastoma multiforme (GBM), is associated with a 12- to 15-month survival after diagnosis despite all forms of treatment, including neurosurgical resection, focal radiotherapy, and conventional chemotherapy. It thus seems that the vast improvements in surgical and adjuvant therapy for cancer over the years have yielded limited clinical benefit in neurooncology. This is largely due to the unique challenges that brain tumors present in terms of their intrinsic biology and anatomical location.

CHALLENGES IN BRAIN TUMOR THERAPY

More specifically, it is the distinct cellular and microenvironmental features within the central nervous system (CNS) that contribute to the refractory nature of brain tumors. These features include:

1. The locally aggressive nature of primary brain tumors within the CNS, which means that microscopic infiltrations of tumor cells into regions of normal brain commonly occur, even in histologically benign disease. Moreover, as secondary brain metastases usually occur via hematogenous spread, the entire

brain can be seeded with micrometastases, even when only a single intracranial lesion is detected. Such microscopic invasion results in high relapse rates, even after apparent complete surgical resection by visual inspection and magnetic resonance imaging (MRI).
2. The genotypic and phenotypic heterogeneity of tumor cells, which enables selection for treatment-resistant cells with redundant pathogenic features, such as immune evasion, angiogenesis, drug resistance, and invasiveness.
3. The presence of the blood–brain barrier (BBB), which regulates the trafficking of substances between the blood stream and the CNS and hinders delivery of many therapeutic agents to the tumor.
4. The distinct molecular environments and cytoarchitecture of different CNS regions, which affect the flow of nutrients, metabolites, cytokines, and neurotransmitters within normal and cancerous tissue, and can interfere with the delivery and efficacy of therapeutic agents.

The cell biology and unique microenvironment of brain tumors necessitate the use of novel technologies to bypass current limitations in treatment, with the ultimate aim of improving patient morbidity and mortality.

THE ASCENT OF NANOTECHNOLOGY

In 1974 Nori Taniguchi coined the term "nanotechnology," which refers to the development of materials and particles with properties that have been specifically designed at the nanometer scale (10^{-9} m). Nanoparticles (NPs) are usually less than 100 nm in size, and can be broadly classified into lipidic NPs, polymeric NPs, and inorganic NPs (Table 7.1).

The application of nanotechnology to medicine involves an extensive degree of collaboration across multiple scientific disciplines, such as microelectronics, surface science, molecular biology, and tissue engineering [7]. The integration of knowledge from these diverse fields ultimately enhances visualization and clinical intervention at the subcellular level. Naturally, this could help address limitations of current imaging and treatment options in brain cancer.

In this chapter we will discuss some current challenges in the management of brain cancer with chemotherapy, radiotherapy, and surgery, and how these challenges can be addressed by nanotechnology. We will subsequently discuss several novel treatment concepts that have been made possible by the unique structural and biochemical properties of nanoparticles.

CHEMOTHERAPY

CROSSING THE BBB

The BBB is often considered the most restrictive barrier in the human body, keeping most small molecules and macromolecules out of the CNS [8]. While the

Table 7.1 Structural and Functional Properties of Different Classes of Nanoparticles

Nanoparticles		Structural and Functional Properties
Lipidic NP	Nanoliposomes	• Bilayered vesicles composed of lipids or phospholipids with aqueous core • Unilamellar or multilamellar • 20–500 nm
	Nanoemulsions	• Oil in water: Oil droplets dispersed in aqueous medium • Water in oil: Water core stabilized by surfactants and cosurfactants • 20–200 nm
	Nanocapsules	• Solid hydrophobic care with outer phospholipid monolayer • 10–200 nm
Polymeric NPs	Micelles	• Hydrophobic core stabilized by hydrophilic shell • 10–150 nm
	Dendrimers	• Spheroid nanostructure consisting of repetitively branched 3D structure • Size depends on number of generations; PAMAM dendrimer has diameter of 1.5–14.5 nm
	Nanogels	• Hydrogel composed of cross-linked ionic and nonionic polymers • Less than 150 nm
	Nanopolymers	• Solid nanopolymers composed of natural or synthetic biodegradable and biocompatible polymers • 10–100 nm
Inorganic NPs	Quantum dots	• Colloidal semiconductor crystals with metalloid crystalline core • Can be coated or conjugated with various molecules • 2–10 nm
	Iron oxide NPs	• Magnetites (Fe_3O_4) & Maghemites (Fe_2O_3) • Superparamagnetic iron oxide (SPIO): 50–150 nm • Ultrasmall SPIO: 10–14 nm
	CNT	• Cylindrical nanostructures made of graphene sheets wrapped onto themselves
	Gold (Au) NPs	• Nanoparticle consisting of gold atoms • Size: ~2.5 nm

NPs, nanoparticles.

selective transport of substances across the BBB is crucial to preventing toxin accumulation in the brain, it also impedes the delivery of palliative chemotherapeutic agents to brain tumors. For instance, although temozolomide (TMZ), a DNA alkylating agent used in the treatment of gliomas [5], is 100% bioavailable

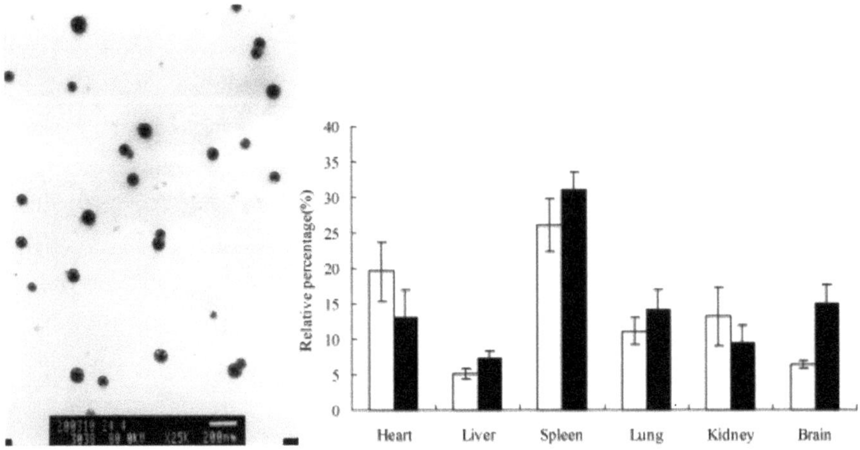

FIGURE 7.3

TMZ solid lipid nanoparticles (TMZ-SLNs) were prepared by an emulsification and low-temperature solidification method (left). Tissue distribution of TMZ after intravenous administration of TMZ-SLNs (*black columns*) and TMZ-Sol (*white columns*) in mice (right).

when taken orally, its concentration in the CNS only reaches approximately 30% of the plasma concentration [9]. Increases in TMZ dosage to increase CNS concentrations are limited by the drug's off-target toxicities, such as cardiac toxicity, bone marrow depression, and oral ulceration.

To address this conundrum, nanoparticles can be used as drug delivery vehicles to increase the localization of administered drugs to the CNS. For instance, TMZ concentrations within the brains of mice were shown to be higher and more sustained following intravenous administration of temozolomide solid lipid nanoparticles (TMZ-SLN) via the tail vein, compared to mice injected with standard TMZ solution (Fig. 7.3). Moreover, the surface functionalization of nanoparticles with tumor-targeting molecules, such as antibodies, peptides (including toxins), and cytokines, can increase the penetrance of the BBB and promote selective uptake into tumors. The use of nanoparticles for specific targeting to tumors would permit escalations in drug doses to increased tumoricidal efficacy, while simultaneously reducing off-target and systemic toxicity [10].

There are several mechanisms by which nanoparticles can be transferred across the BBB. The first is adsorptive-mediated transcytosis, which exploits the electrostatic interactions between the positively charged moieties of nanoparticles and negatively charged brain endothelial cells membranes to move across the BBB [11]. For example, cationic albumin-conjugated polyethylene glycol (PEG)-coated nanoparticles (CBSA-NP-ACL) have been loaded with aclarubicin and administered through intravenous injection to rats with intracranially implanted C6 glioma cells. One hour postinjection, a 6-coumarin fluorescent label showed that CBSA-NPs had accumulated in the tumor mass to a greater extent than nanoparticles without conjugated CBSA, suggesting that the cationic charge could significantly improve nanoparticle localization to the brain [12]. However, the specificity of this targeting strategy is poor since cationic moieties can interact with any negatively charged cell membrane constituent [11].

Alternatively, transporter-mediated transcytosis exploits the presence of certain transporter molecules on the cerebral endothelium to deliver drugs that closely mimic the endogenous substrates. For instance, a novel glycosyl derivative of cholesterol was developed as part of a liposome-based drug delivery system to the brain, as the glycosyl group could target the glucose GLUT transporters found in high concentrations on the BBB [13]. Yet another mechanism of transport across the BBB is receptor-mediated transcytosis, in which nanoparticles are engineered to express surface ligands for receptors that are highly expressed on brain capillary endothelial cells. In this vein, studies suggest that the attachment of apolipoprotein E to the surface of nanoparticles facilitates drug transport across the BBB, presumably following interaction with lipoprotein receptors on the brain capillary endothelium [14,15].

After crossing the BBB, nanoparticles still have to selectively accumulate within tumor tissue. In contrast to small molecules that diffuse freely into and out of tissues, nanoparticles behave as macromolecules and therefore tend to selectively accumulate within tumors via the enhanced permeability and retention (EPR) effect. The EPR effect refers to the phenomenon in which high-molecular-mass molecules and liposomes (MW $\sim$45 kDa) accumulate in cancer tissue as a consequence of active angiogenesis, the expression of vascular mediators of extravasation (e.g., nitric oxide, vascular endothelial growth factor, and bradykinin), and the development of immature or malformed blood vessels containing wider fenestrations. These changes to vascular architecture and permeability are accompanied by the loss of effective lymphatic drainage pathways, leading to the decreased clearance of nanoparticles from tumors. Studies suggest that nanoparticles smaller than 100 nm more effectively extravasate from leaky and disorganized blood vessels within tumors, while nanoparticles larger than 20 nm are more effectively retained within the tumor [16]. Overall, the EPR effect results in the retention of nanoparticles within tumor tissue long after their serum levels decline, making nanoparticles a useful platform for drug delivery.

SELECTIVELY TARGETING CANCER CELLS

Unfortunately, a common effect of current cytotoxic drugs is damage to surrounding normal brain tissue. The small size of nanoparticles addresses this problem of nonspecific cellular destruction by permitting greater precision in cell targeting, while also allowing access to tumors in deep or anatomically challenging regions of the body [7].

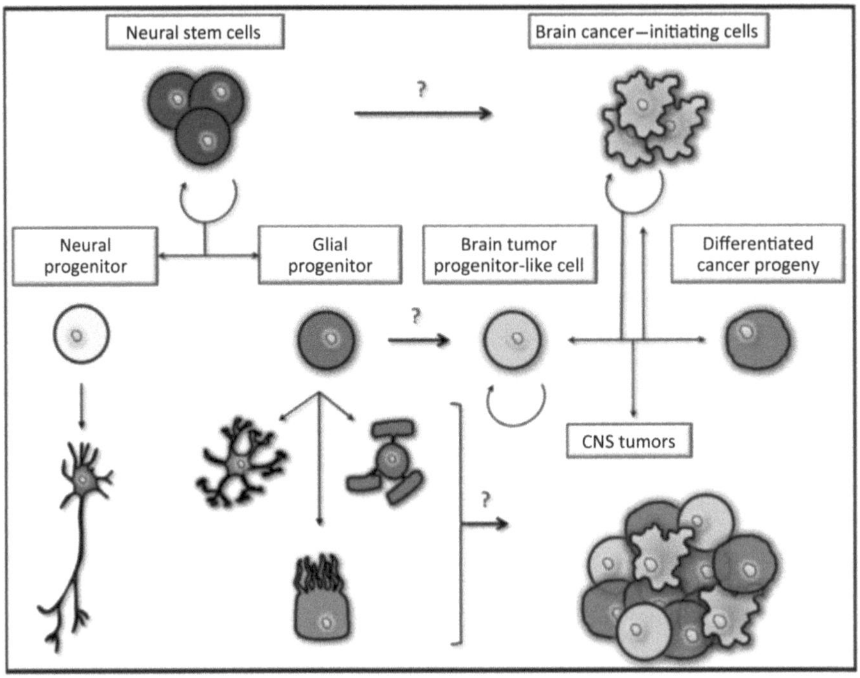

FIGURE 7.4

Relationship between the differentiation of neural stem cells (NSCs) and tumor formation by a brain cancer–initiating cell. NSCs differentiate into glial or neural progenitors. Glial progenitors give rise to oligodendrocytes, ependymal cells and astrocytes, while neural progenitors give rise to neurons. Brain tumors may develop following the transformation of NSCs into brain cancer–initiating cells, which in turn give rise to more differentiated tumor cells. Alternatively, glial progenitors could induce tumors following malignant transformation of normal progenitor cells (brain tumor progenitor-like cells). Mature cells in the brain such as neurons, oligodendrocytes, ependymal cells, and astrocytes have traditionally been seen as potential players in brain cancer tumorigenesis [19].

Copyright © 2013 MDPI AG. Reproduced with permission from Goffart N, Kroonen J, Rogister B. Glioblastoma-initiating cells: relationship with neural stem cells and the micro-environment. Cancers (Basel). 2013;5(3):1049–71.

Recently, it has been suggested that cancer therapies for both solid organ and hematopoietic neoplasms must specifically target a subpopulation of tumor cells that are responsible for initiating and maintaining the tumor. These cells share similar biological properties to normal adult stem cells, and are known as cancer-initiating cells (Fig. 7.4). In 2004 it was reported that the $CD133^{+}$ fraction of cells from human brain tumors exhibited stem cell properties in vitro, and showed a significant capacity for self-renewal and exact recapitulation of original tumors in xenograft assays [17]. These brain cancer-initiating cells (BCICs) form a small part of the tumor, but are able to initiate and repopulate a tumor, even if most of it has been removed. BCICs are refractory to many conventional anticancer treatments, and likely contribute to the high recurrence rate of brain tumors [16].

The use of nanotechnology to specifically target these BCICs, as opposed to the bulk of the tumor that is composed of nontumorigenic cells, may help reduce the high relapse rates currently observed in the clinic. For example, nanoparticles can be designed to target surface markers that are highly expressed on BCICs. An example of such a maker is a 200- to 240-kDa intermediate filament protein known as nestin, which is commonly used to isolate and study BCICs. A nestin-targeted peptide, referred to as AQYLNPS, has been developed through in vitro phage display technology and showed selective binding to undifferentiated glioma-initiating cells [18]. Such BCIC-targeted peptides may be an important component of future drug-nanoparticle conjugates for brain tumors. However, steps must be taken to ensure that neural stem cells (NSCs), which share many characteristics and surface markers with BCICs, are not excessively damaged by BCIC-directed therapy.

CONTROLLED DRUG RELEASE

Another significant advantage of nanoparticles in chemotherapy is their ability to protect drugs against biotransformation and rapid clearance in vivo. Controlled release drug delivery systems can also be used to ensure a slow and prolonged release of drugs over time. By altering the pharmacokinetic profile of drugs in this way, nanoparticles allow a similar drug dose to yield a lower but more sustained drug concentration in the body, which could lower systemic toxicity and enhance antitumor effects. Microreservoir drug-loaded arrays that facilitate the controlled release of chemotherapeutic drugs are thus being intensively studied (Fig. 7.5) [20,21]. A resorbable, multireservoir polymer microchip loaded with 1,3-bis(2-chloroethyl)-1-nitrosourea (BCNU) has been tested in a syngeneic Fischer 344 9L gliosarcoma rat model. BCNU is bifunctional alkylating agent of the nitrogen mustard class of chemotherapeutic agents and induces cell death by forming interstrand cross-links in DNA, thereby preventing DNA replication and transcription. The polymer microchip increased BCNU stability at 37°C, and the successfully decreased tumor size in a manner that was dependent on the concentration of BCNU loaded onto the microchip [21].

Drug delivery systems can also be designed to release pulses of different drugs at stipulated time intervals. As drug release depends on the rupture of the membranes covering the microchip reservoirs, variation in the timing of drug release can be achieved with the use of materials with different molecular masses in the membrane covering of the drug reservoirs. Future development of this technology might even allow for drug release to be controlled by a computer-based, wireless programing device.

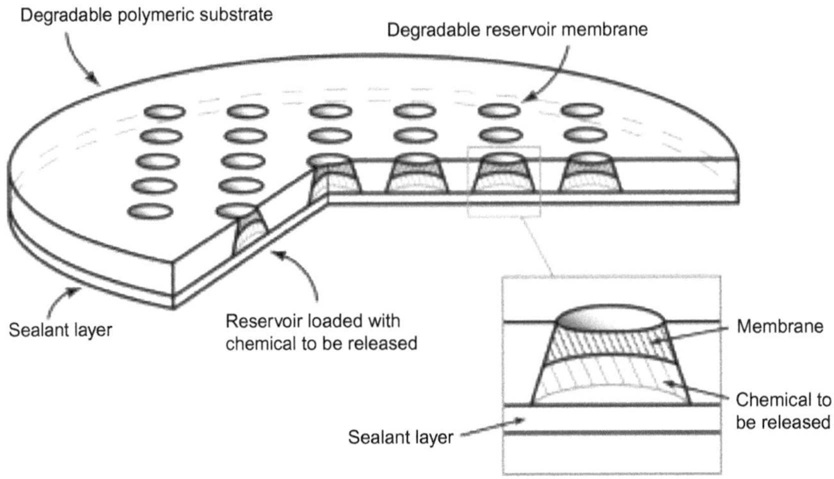

FIGURE 7.5

Schematic of a drug-releasing polymer microchip. Cytotoxic drugs such as BCNU are loaded into the reservoirs, and degradation of the reservoir membranes releases the drug into the body [20].

It is worth noting that controlled drug release devices have been tested in humans for the treatment of osteoporosis. The pulsatile release of parathyroid hormone via these devices has shown both safety and efficacy, such that this technology may indeed play a significant role in the design of future chemotherapy regimens.

OVERCOMING MULTIDRUG RESISTANCE

The genetic heterogeneity of tumor cells means that therapy selects for resistant phenotypes. Treatment can thus lead to the development of tumors that are resistant to multiple anticancer drugs [22]. One of the major mechanisms of multidrug resistance involves overexpression of the p-glycoprotein (pgp) efflux transporter, which efficiently removes anticancer agents from tumor cells. For instance, resistance to paclitaxel treatment is partially attributed to reduced paclitaxel concentrations in brain tumor cells as a result of pgp-mediated efflux. To overcome this form of multidrug resistance, drugs can be associated with nanoparticles to circumvent efflux mechanisms, accumulate intracellularly, and enhance cytotoxicity. This concept was tested in an in situ rat brain perfusion model, where paclitaxel was entrapped in novel acetyl alcohol/polysorbate nanoparticles (Fig. 7.6). Entrapment of paclitaxel in nanoparticles significantly increased brain uptake and increase paclitaxel toxicity in pgp-expressing tumor cells. It was suggested that encapsulation of paclitaxel within nanoparticles might interfere with drug

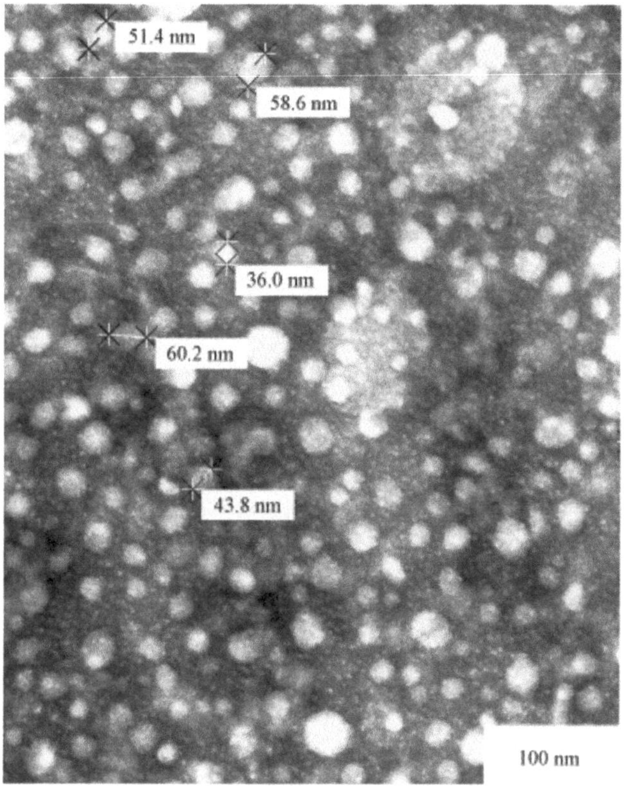

FIGURE 7.6

Nanoparticle formulations containing paclitaxel were produced with oil-in-water microemulsion templates, and their size and morphology confirmed by TEM [23].

binding to pgp, thereby facilitating tumor cell accumulation of the otherwise effluxed drug [23].

While the intrinsic material properties of nanoparticles are able to cripple certain mechanisms of multidrug resistance, tumor cells can also be sensitized to conventional drugs through the rational design of nanoparticle structure. For example, some tumor cells acquire resistance to therapy with cytotoxic agents by upregulating their intrinsic DNA repair machinery. For example, high expression of the DNA repair enzyme O^6-methylguanine methyltransferase (MGMT) can reduce tumor cell sensitivity to cytotoxic anticancer agents by reversing the DNA damage induced by this these drugs. Administration of O^6-benzylguanine (BG), which can directly deplete MGMT and thus inhibit repair, once again renders tumors susceptible to cytotoxic therapy. To optimize the effects of BG, ingeniously designed chitosan surface-modified poly(lactide-*co*-glycolides)

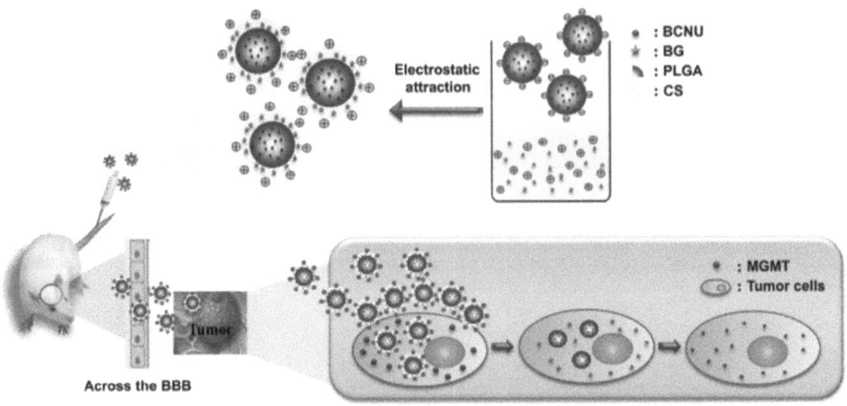

FIGURE 7.7

The preparation and in vivo effects of cationic core–shell PLGA nanoparticles loaded with BCNU and BG. Nanoparticles were constructed based on the electrostatic attraction between PLGA and CS, and prepared using the emulsion-solvent evaporation method. Following intravenous injection, nanoparticles moved across the BBB via adsorptive-mediated endocytosis and accumulated in tumor via the EPR effect, where they consumed MGMT by releasing BG prior to BCNU [24].

(PLGA/CS) nanoparticles were loaded with BCNU as well as BG. Crucially, the core of the nanoparticles consisted of PLGA carrying BCNU, while the external coating shell consisted of chitosan (CS) loaded with BG. The presence of BG in the external shell layer meant that the drug could be released from the nanoparticles before BCNU, to consume MGMT in the tumor cells and enhance the therapeutic effect of BCNU (Fig. 7.7). In fact, BCNU/BG nanoparticles were 3.5× more cytotoxic to F98 glioma cells than free BCNU/BG. In addition, F98 cells with higher MGMT levels were shown to be more sensitive to the treatment than C6 cells, which have lower MGMT levels [24]. As low expression of MGMT correlates with increased survival following chemotherapy [25–28], subtle but informed changes to nanoparticle design could indeed achieve clinically significant benefits, by facilitating subversion of tumor drug resistance mechanisms.

RADIOTHERAPY

TARGETING RADIATION DAMAGE TO TUMORS

Radiotherapy is another important standard-of-care therapy for patients with brain tumors. Not unlike chemotherapy, the cumulative dose of radiation that can be safely administered to a patient is limited by off-target destruction of normal tissue. Internal radiotherapy, in which radioisotopes or radionuclides are administered directly to patients, provides a means for radiation-induced cell damage to be targeted to tumors. These radionuclides are preferentially taken up by tumor cells, and thus simultaneously increase the radiation dose to which the tumor is exposed, while localizing damage to tumor cells. The homing of radionuclides to tumors can be further improved through radioimmunotherapeutic approaches, in which radionuclides are conjugated to monoclonal antibodies and peptides that target tumor-specific cell receptors. However, the relatively low and variable expression of these receptors on tumor cells mean that even radioummunotherapeutic targeting strategies are insufficient to guarantee localization of toxic radionuclides to tumors.

Nanoparticles can once again address the challenge of localizing radiation damage to tumor cells by serving as delivery vehicles for radionuclides. For example, dosimetry and survival studies in a rat model suggest that systemically administered liposomal formulations of ^{188}Rhenium can be passively targeted to gliomas. The increase in the life span of treated rats, coupled with the safety of the systemically administered radioisotopes, suggest that nanoparticles could represent an important step forward in the treatment of brain malignancies with internal radiotherapy [29]. Moreover, nanoparticle-conjugated radionuclides could be designed to enable the simultaneous visualization and radioablation of tumor cells. This administration of therapy at the time of diagnosis exemplifies the "theranostic" approach to cancer that has received growing attention in recent years. Briefly, theranostics involves the seamless integration of diagnosis and therapy into a single step. It has been made an increasingly realistic prospect in the clinic due to the recognition unique combination of diagnostic and therapeutic capabilities that nanoparticles offer. In this way, nanoparticles could well form the basis of a new paradigm in cancer medicine [30].

Yet, notwithstanding efforts to improve the specificity of radiation damage and facilitate efficient treatment upon diagnosis, the ultimate success of radiotherapy depends on our ability to overcome the resistance of certain brain tumor cells to ionizing radiation [31]. In a manner analogous to the selection of drug-resistant tumor cells during chemotherapy, radiotherapy can select for radiotherapy-resistant tumor cells, which are responsible for the relapse of radiotherapy-resistant tumors. While higher radiation doses could help kill a greater extent of these initially radioresistant tumor cells, they would also exacerbate damage to normal surrounding tissue. Consequently, an innovative method to sensitize the radioresistant fraction of tumors to standard radiation doses is necessary.

One recently proposed method to achieve this involves the administration of PEGylated-gold nanoparticles (GNPs) with radiotherapy. GNPs were shown to sensitize tumor cells to radiation by increasing radiation-induced cellular DNA damage and reduced clonogenic survival in human GBM-derived cell lines, with a dose-enhancement ratio of approximately 1.3. Increased DNA damage was also observed in vitro in brain-derived endothelial cells, suggesting that GNPs could also cut off the supply of nutrients to tumors in vivo by sensitizing tumor-associated blood vessels to radiation [32]. The administration of radiotherapy to GNP-treated mice also increased survival in mice with orthotopic GBM tumors. Interestingly, pretreatment of mice with radiotherapy prior to GNP administration could increase extravasation and deposition of GNPs in brain tumors [32]. These data thus implicate GNPs as a promising radiosensitizing agent for brain tumor therapy, and further suggest that optimizing the temporal administration of radiotherapy and radiosensitizing nanoparticles could lead to a strong synergistic effect on tumor destruction [32,33].

REPAIR OF RADIATION-INDUCED BRAIN DAMAGE

Despite impressive efforts to preferentially target and radiosensitize tumor cells, a certain degree of off-target radiation damage to normal tissue may be unavoidable. NSCs and neural progenitor cells are thus being explored as a means to repair inevitable radiation-induced brain damage. These self-renewing, multipotent cells can be induced to proliferate and differentiate, either in vivo or in vitro, into terminally differentiated neurons and glia. In the adult mammalian brain, the subventricular zone (SVZ) of the hippocampus is the main neurogenic niche, and NSCs within it can eventually give rise to new neurons, astrocytes, and oligodendrocytes. Nanoparticles could hence be used to initiate neuroregeneration following radiotherapy by promoting the proliferation and differentiation of these NSCs. For instance, polymeric nanoparticles loaded with retinoic acid were shown to induce *in vivo* differentiation of NSCs in the SVZ in mice. The intracellular delivery of retinoic acid by nanoparticles could activate nuclear retinoic acid receptors in NSCs, decrease stemness, and upregulate expression of pro-neurogenic genes such as Mash1 and Neurogenin1 that may be required for the successful commitment of NSCs in the SVZ niche to the neuronal progenitor lineage [34].

Crucially, NSCs appear capable of migrating toward lesions within the brain, meaning that they could specifically repopulate areas that have sustained damage following radiotherapy. However, despite their exciting neurogenerative capabilities, pure NSC populations are difficult to isolate. One study thus investigated the neuroregenerative capacity of rat mesenchymal stem cells (rMSCs) labeled with 1-hydroxyethylidene-1,1-bisphosphonic acid–coated superparamagnetic iron oxide (SPIO) nanoparticles. Interestingly, iron-labeled rMSCs in the SVZ were able to differentiate and migrate when a mechanical lesion of the olfactory bulb was performed. The long migration distance from the SVZ to the olfactory bulb through the rostral migratory stream was highly suggestive of a potent migratory activity of iron-labeled rMSCs, which could be induced only in the presence of damage. Further studies on the precise intra- and extracellular cues for differentiation and migration are necessary to fully determine the role that mesenchymal stem cells (MSCs) could play in the repair of radiation-induced damage [35].

Alternatively, as cellular differentiation is more difficult to manipulate and monitor in vivo, stem cells can be cultured under specific in vitro conditions to

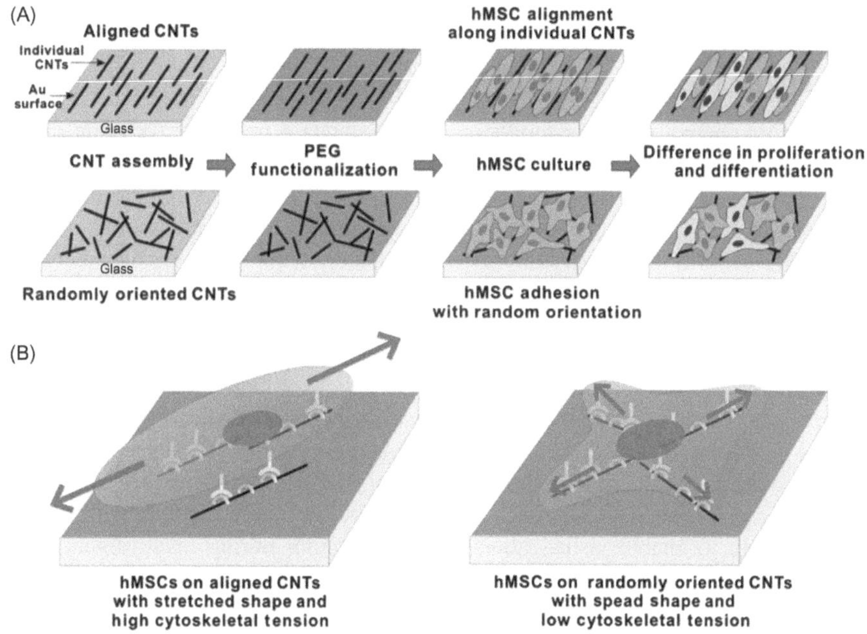

FIGURE 7.8

Growth and differentiation of hMSCs on CNT networks with the different arrangement of individual CNTs. (A) CNTs were assembled on Au substrates with a low surface density in an aligned or a randomly oriented formation. Bare surface regions between the CNTs were functionalized with thiolated PEG. hMSCs were cultured on the substrates and showed different growth and differentiation patters. (B) Proposed model to explain the hMSC response to aligned and randomly oriented CNT networks. The hMSCs on aligned CNT networks elongated along the direction of the CNTs because of a high affinity between CNTs and cells. High cytoskeletal tension was induced in the stretched hMSCs on aligned CNT networks [36].

generate transplantable neural tissue. Stem cell fate can be effectively controlled in vitro by optimizing cell culture properties and using engineered scaffolds to provide dynamic microenvironmental and morphological cues for cell proliferation and differentiation at the nanoscale. Carbon nanotubes (CNTs) have emerged as a promising new cell culture substrate to accomplish this due to the degree of control that they afford over stem cell differentiation. CNT scaffolds functionalized with thiolated PEG have been successfully used to promote neuronal differentiation of human MSCs (Fig. 7.8) [36], and could thus act as potential supporting matrices for stem cell–derived neuronal transplants.

SURGERY

INTRAOPERATIVE DELINEATION OF TUMORS

Even before chemotherapy or radiotherapy, the standard treatment for brain tumor patients usually entails maximal surgical resection. In fact, the completeness of tumor resection during surgery is one of the most important determinants of patient prognosis and disease progression, across all age groups, tumor grades, and sites of origin [37,38]. This is because tumors most frequently recur at the interface between tumor cells and normal brain. However, the extent of tumor resection must be balanced by the need to minimize the removal of normal brain tissue, particularly when dealing with areas that are close to important neuroanatomical pathways. A method to reliably and accurately identify malignant cancer cells at the tumor margins would help mitigate the risk of unnecessarily removing healthy tissue.

The neurosurgeon's ability to discriminate between normal and cancerous tissue has improved substantially with the development of methods for intraoperative optical tumor delineation. This approach was pioneered by Moore in 1948, with the use of fluorescein to enable visualization of a brain tumor during neurosurgery [39]. Various methods since then have utilized compounds such as 5-aminolevulinic acid (5-ALA), which induces synthesis and accumulation of fluorescent protoporphyrin IX in malignant glioma tissue, to facilitate intraoperative recognition of cancerous brain tissue. Recently, there is growing evidence that the superior targeting ability of nanoparticles, as well as their susceptibility to external manipulation, may further enhance the optical delineation of tumors during surgery.

Iron oxide nanoparticles have been widely studied as intraoperative MRI contrast agents, as they cause enhancement in brain tumors following phagocytosis by cancer cells over a much longer period of time (24–48 hours), compared to conventional contrast agents. In fact, a single dose of iron oxide nanoparticles administered 24 hours prior to surgery is sufficient to provide intraoperative contrast enhancement, without concomitant enhancement of surgically manipulated tissue. The ability to avoid enhancement in surgically manipulated tissue is an important benefit from a practical point of view—it means that neurosurgeons can confidently remove all enhancing tissues detected, as these tissues are highly likely to be cancerous.

However, the imaging options discussed thus far are generally contingent upon the use of bulky, specialized equipment for tumor visualization. Continued progress in nanotechnology may eventually permit visualization of cancerous tissue without the need for external equipment. To this end, a methylene blue–loaded polyacrylamide nanoparticle named Nanocyan has been investigated for its ability to clearly delineate neoplastic tissue, without additional equipment, under the normal lighting conditions of an operating theater. The surface of Nanocyan is also functionalized with F3 targeting peptides and a coating of cysteine to prevent nonspecific binding. It has been shown to deeply stain 9L gliosarcoma cells in an F3-dependent manner in in vitro studies. Importantly, the components of Nanocyan are nontoxic. The relevance of Nanocyan for future clinical use will be evaluated as more in vivo data becomes available [40].

Another promising suggestion to enable intraoperative optical delineation of tumors without the assistance of bulky or immobile equipment involves the use of a handheld Raman spectrometer to visualize tumor cells that have taken up nanoparticles with unique spectral characteristics, known as a Raman "fingerprint." For instance, gold silica surface-enhanced Raman scattering (SERS) nanoparticles, which were intravenously administered to mice with GBM, enabled the accurate demarcation of tumor cells with the use of a handheld Raman device. The GBM could be resected more completely with this device than with a static Raman microscope, as the handheld probe could be used at various angles following resection steps to ensure thorough tumor removal. In this way, the handheld Raman scanner introduces a degree of maneuverability that is not possible with current intraoperative imaging modalities. SERS-image-guided tumor resection provides additional advantages over imaging techniques. For instance, the rival technique, which detects fluorescence derived from 5-ALA, is limited by the presence of a certain degree of background autofluorescence in normal brain tissue [41]; this can lead to the excision of healthy tissue. Similarly, gadolinium (Gd)-based agents diffuse over time from initial tumor into the edematous peritumoral area, resulting in an inaccurate estimation of the tumor margin [42]. In contrast, the unique Raman fingerprint generated by SERS nanoparticles reduces the risk of a false-positive signal [42]. In addition to this increased specificity, the Raman fingerprint of SERS nanoparticles provides a more prolonged optical signal than conventional organic dye molecules, as the latter are susceptible to photobleaching.

Further suggestions have been put forward to improve the targeting of SERS nanoparticles to brain tumors. For instance, surface functionalization of nanoparticles with the anti-EGFR receptor panitumumab could enhance their interaction with and internalization by tumor cells with a high surface expression of EGFR. Following administration of functionalized nanoparticles, magnetic resonance image–guided transcranial focused ultrasound can be used to reversibly disrupt the BBB and increase penetration of nanoparticles into the brain. A study using this two-step targeting procedure showed that functionalized SERS nanoparticles preferentially accumulated in EGFR-positive GBM cells at the leading edge of the tumor. As GBMs usually recur at this interface between tumor cells and normal brain, the reliable and accurate targeting strategy could help to reliably and accurately demarcate tumor tissue, and in turn reduce relapse rates. Future progress in thus area may even include the addition of a theranostic component to the handheld Raman scanner to facilitate the simultaneous detection and destruction of SERS-positive tumor cells, possibly with the use of an ultrasonic aspirator or laser.

NANOELECTROMECHANICAL SYSTEMS

Nanoelectromechanical systems (NEMSs) are devices that integrate electrical and mechanical functions at the nanoscale. They consist of miniaturized electrical and mechanical apparatuses such as actuators, beams, sensors, pumps, resonators, and motors. These components convert one form of energy into another, which can be quickly and conveniently measured. These devices can function as biosensors to monitor important physiological variables during surgical procedures, such as intracranial pressure, cerebrospinal fluid (CSF) pulsatility, weight load, and strain.

NEMSs provide three main advantages as mechanical biosensors in surgery. First, they can achieve mass resolution at the nanogram scale when operating in a fluid environment, as the minimum detectable mass added is proportional to the total mass of the device. Second, the ability of an NEMS device to be displaced or deformed—known as mechanical compliance—increases with uniform reduction of its dimensions. This high degree of mechanical compliance allows an applied force to be translated into a measurable displacement, such that even the miniscule forces governing cellular and subcellular interactions can be quantified. For example, NEMS sensors can resolve forces as small as 10 pN, making them sensitive enough to detect the breaking of hydrogen bonds. Third, small fluidic mechanical devices can exhibit fast response times, which would facilitate real-time monitoring of biological processes [43].

An implantable bioresorbable nanoporous silicon device, with dimensions of 1 mm × 2 mm × 0.08 mm, was found to provide a reliable assessment of intracranial pressure in rats (Fig. 7.9) [44]. The resistance of the sensing element increased monotonically in a linear manner across the full range of pressures (0–70 mm Hg) that are relevant to intracranial monitoring. The device was also

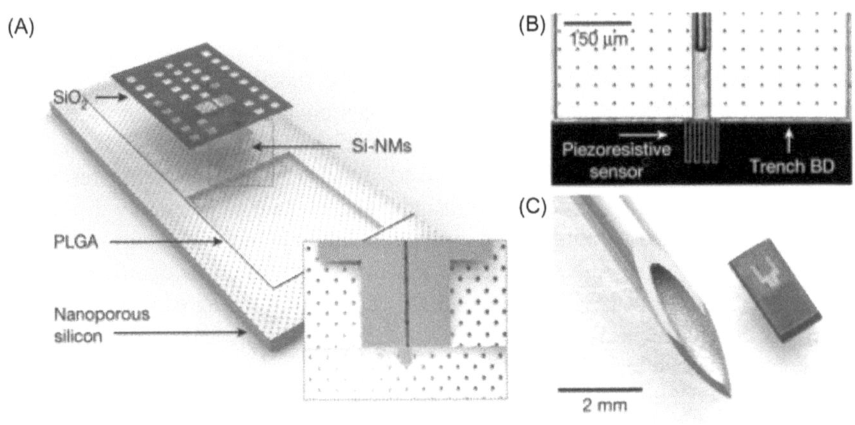

FIGURE 7.9

(A) Schematic of biodegradable NEMS pressure sensor. Inset shows location of the silicon-nanomembrane strain–gauge. (B) Optical micrograph of the strain–gauge region. (C) Image of complete device [44].

Copyright © 2016 Nature Publishing Group. Reproduced with permission from Kang SK, Murphy RK, Hwang SW, Lee SM, Harburg DV, Krueger NA, et al. Bioresorbable silicon electronic sensors for the brain. Nature 2016;530(7588):71–6.

amenable to wireless transmission of information, and could therefore be used to monitor neurophysiological variables even after surgery. In vivo tests using the silicon device to measure intracranial pressure compared favorably with current methods that rely on wired sensors and are thus not suitable for implantation and postsurgical monitoring [45]. Furthermore the device was seen to dissolve over time when exposed to biofluids, such as CSF. As only biocompatible end products were eventually formed, subsequent invasive procedures to remove implanted NEMS devices could be rendered unnecessary in future clinical settings.

The implantable nature of these devices has significant implications for the postsurgical follow-up of brain tumor patients. For example, implanted sensors embedded in the resection cavity could facilitate a prompter detection of tumor recurrence, compared to the current strategy that depends on interval MRI. Sensor arrays could thus be designed to register changes in tissue impedance, hypoxia, pH, or temperature to identify the hallmark signs of tumor progression. This early warning system would allow proactive rather than reactive initiation of secondary therapies. Furthermore the integration of miniaturized sensor arrays with an NEMS component to destroy adjacent tissue could enable the immediate in situ ablation of recurring tumors. The administration of local therapies through this neurally embedded system (e.g., hyperthermia induced by passing a current between two electrodes, ultrasound or UV light, or release of an aliquot of chemotherapy) could minimize the side effects of systemically administered therapies. Of course, the introduction of foreign bodies such as NEMS devices into the brain is inevitably associated with a certain degree of parenchymal damage and local neuronal death, along with risks of bleeding, infection, and seizures. Foreign bodies can also cause the activation of microglia and astrocytes and reactive gliosis, which in turn can hinder the function of implanted NEMS devices. Future work will thus need to look into ways to improve the biocompatibility and safety of implantable devices.

Apart from their potential use in neuromonitoring, NEMS technology is also integral to the development of nanotools that could be used to perform intricate nanosurgeries within the CNS. For instance, a recently developed nanoknife has been successfully used to cut individual axons of peripheral nerves in an in vivo mouse model. Such technology could eventually permit a more precise mechanical disconnection of individual white matter bundles during neurosurgical resection, and may lead to a significant improvement in surgical outcome for brain tumor patients. In fact, avenues for intervention at the cellular and subcellular level during neurosurgery could be made possible with NEMS technology. For instance, nanowires could be integrated with cellular components to create a direct bridge between the cell and the external environment within the control of neurosurgeons, in order to facilitate the delivery of biological compounds. The safety and efficacy of nanowire technology was shown in a study in which atomic force microscopy tips were repurposed for the delivery of fluorescent nanoparticles [46]. The tip diameter used was less than 10 nm, as tip lengths substantially smaller than the cell can mitigate physical damage to the cells. With further development, NEMS technology could thus provide neurosurgeons with an unprecedented level of control over the cellular environment within the brains of cancer patients.

NEUROIMAGING

Neuroimaging through serial MRI remains the primary monitoring tool for patients diagnosed with glioblastoma, with assessments performed every 2–3 months during treatment. However, standard MRI contrast studies are sometimes misleading, even when adhering to the Macdonald criteria. Typical MRI modalities, such as T2, T1 with Gd, and fluid-attenuated inversion recovery, remain insufficient to confidently differentiate between recurrence and radionecrosis or pseudoprogression during the first few months following completion of radiotherapy and TMZ. Both radionecrosis and pseudoprogression appear as masses mimicking progression. However, radionecrosis is in actuality a complication of radiotherapy that may develop months to years following radiotherapy as a consequence of off-target damage to normal brain tissue. Pseudoprogression is a transient period of apparent radiographic deterioration, sometimes associated with neurological symptoms, that resolves spontaneously. Thus, instead of reflecting an actual increase in tumor size, radionecrosis, and pseudoprogresison are consistent with the increased tumor cell killing and normal tissue responses to chemoradiotherapy, including BBB breakdown, ischemia, and inflammation. However, MRI features such as the "Swiss cheese" or "soap bubble" patterns have failed to consistently identify necrosis over progression in different studies [47,48]. Moreover, pseudoprogression, observed in 20–30% of patients of receiving standard radiotherapy and TMZ, can currently only be proven post hoc if no treatment is administered and the lesion subsequently regresses.

The limitations of current MRI imaging modalities can be addressed by the unique paramagnetic properties of magnetic nanoparticles (MNPs). For example, ultrasmall SPIOs (USPIOs), which are smaller than 50 nm, have become increasingly promising as T1 and T2 MRI contrast agents. They can be visualized as a hypointense (dark) signal in T2-weighted MRI sequences, or as a hyperintense (bright) signal in T1-weighted MRI sequences. They are associated with fewer adverse renal effects and longer contrast enhancement than Gd-based contrast agents, which are rapidly eliminated by the kidneys. In fact, USPIOs can provide a peak contrast enhancement within brain tumors at about 24–28 hours and can persist for up to 72 hours after administration. Importantly the USPIO ferumoxytol has been investigated recently for its superior sensitivity to Gd-based agents in distinguishing pseudoprogression from true tumor progression with the use of DSC-MRI. Briefly, DSC-MRI is used to model relative cerebral blood volume (rCBV) in patients, with a high rCBV indicating active neovascularization and tumor progression. Currently, this approach relies on intravascular localization of Gd-based contrast agent in the brain. However, rapid extravasation of Gd-based agents from blood vessels into the extravascular space can result in falsely low rCBV in some patients with true progression. As a blood pool agent with a larger size than Gd agents, ferumoxytol is confined largely to the intravascular space. Its long half-life of about 12 hours also makes steady-state CBV mapping possible. A pilot study on 14 patients with GBM found that DSC-MRI with ferumoxytol provided a more accurate estimation of tumor rCBV than DSC-MRI with gadoteridol. Specifically the study found that lesions demonstrating increased enhancement on T1-weighted MRI as well as a low ferumoxytol rCBV are likely exhibiting pseudoprogression. On the other hand, increasing enhancement with high ferumoxytol rCBV was suggestive of active tumor progression. These exciting observations certainly warrant further evaluation of the potential use of USPIOs to differentiate between tumor progression and pseudoprogression [49].

Additionally the remarkable mechanical properties of optical semiconductor nanocrystals known as quantum dots are being increasingly acknowledged in the field of cancer imaging. The prototypical quantum dot is composed of a cadmium selenide core surrounded by a zinc sulfide shell. These quantum dots can then be functionalized of with conjugated fluorescent proteins to provide several important advantages over conventional fluorophore-based visualization techniques. For instance, in addition to undergoing minimal photobleaching and showing greater stability than typical fluorophores, quantum dots exhibit dramatically enhanced signal detection. Their broad absorption spectra but narrow emission spectra manifests as a high extinction coefficient for a comparable quantum yield, a greatly increased signal-to-noise ratio, and hence a much brighter signal [50–53] Estimates suggest that this impressive combination of physical and optical factors makes quantum dots 20× brighter and 100× more stable than traditional fluorescence vectors [54].

Quantum dot labeling has already found broad application in vitro for diverse cell types, and shows profound utility in single-particle tracking of targets within live cells [55–57]. However, despite widespread contemporary usage, advances in the labeling of neural tissue have been slower to progress. Care must be taken not to fall prey to the tacit assumption that what applies to other cell types automatically applies to neurons and glia. Moreover, while quantum dot technology exerts minimal in vitro cytotoxicity, in vivo applications present a far more dynamic challenge in terms of cytocompatibility and system reactivity—especially since long-term experiments have shown that these quantum dots can remain detectable in mice for after at least 4 months [58].

NOVEL THERAPIES

PHOTODYNAMIC THERAPY

While chemotherapy, radiotherapy, and surgery form the current standard of care for patients with brain tumors, the development of nanotechnology has opened up new avenues of treatment. One of the most exciting of these is photodynamic therapy (PDT), in which patients are injected with a photosensitizing agent that is preferentially retained within cancer cells. After 24–72 hours, when most of the photosensitizer has left normal cells but remains in cancer cells, the tumor is exposed to a specific wavelength of light. Appropriate photosensitizers have a high extinction coefficient in the far-red or near-infrared (NIR) spectral region and a high yield of the long-lived triplet electronic state, formed from the excited singlet state by intersystem crossing. Thus the photosensitizing agent in the tumor absorbs the light, which results in the generation of highly toxic reactive oxygen species (ROS) that destroy tumor cells. PDT is also thought to activate antitumor immunity and to damage tumor-associated blood vessels, thereby compromising its supply of nutrients. Presently, PDT is usually undertaken during surgical resection procedures. Repeated application of light following surgery is occasionally undertaken, with access maintained through the skull.

One nanomaterial that has been extensively studied for the purposes of PDT is graphene, a carbon allotrope with a bidimensional hexagonal structure composed of a single layer of sp2-bound carbon atoms. The ultrahigh surface area afforded by its polyaromatic structure and high optical absorption in the NIR region make it particularly promising for photoablation of tumors [59]. Akhavan et al. have recently studied developed reduced GO nanomeshes (rGONMs) in an in vivo mouse model. They found that injection of just 10 μg/ml of rGO-based composites, together with ultralow laser irradiation, achieved complete tumor ablation. The higher efficiency of rGONMs, compared to rGO sheets and reduced graphene oxide nanoplatelets (rGONPs), was mainly attributed to the superior NIR absorbance of the former [60]. Furthermore, fluorescence imaging revealed that rGONMs functionalized with arginine–glycine–aspartic acid (RGD), which targets αβ integrin receptors overexpressed on U87 glioma cells, showed better tumor localization. These results suggest that nanocomposites such rGONM-PEG-Cy7-RGD could be instrumental to the development of cancer therapies with low side effects [60].

However, the light needed to activate most photosensitizers is unable to penetrate more than 1 cm of tissue. Absorption of light by melanin in the skin also limits the amount of light energy that can pass beyond the skin. Consequently, PDT is usually used to treat superficial tumors on or just under the skin, or on the lining of internal organs or cavities. PDT is also less effective in treating large tumors, as light cannot penetrate the core of these tumors. The exceptional optical properties of certain classes of nanomaterials offer a means to circumvent this problem. One type of nanomaterial known as upconversion nanoparticles (UCNPs) serve as nanotransducers that convert deeply penetrating NIR light, which does not have enough energy to promote the triplet to singlet state conversion in molecular oxygen, into visible light (VIS). UCNPs participate in a nonlinear optical process that converts two or more low-energy pump photons from the NIR spectral region (700–1100 nm) to a higher-energy output photon with a shorter wavelength. Shorter wavelength VIS light is then transferred to a PS with an excitation band matching the emission of the UCNPs. Excitation of UCNPs with NIR light thus triggers fluorescence resonance energy transfer to the attached PS, ultimately resulting in the generation of cytotoxic ROS that can destroy tumor cells.

THERMOTHERAPY

Thermotherapy is another form of therapy made possible by the development of nanotechnology. It exploits the fact that moderate elevation of targeted areas of the body above 40°C can damage proteins and structures within cells, resulting in apoptotic cell death. However, heat applied to the head is shielded by the skull, which results in a less-than optimal temperature increase in the brain. At the same time, temperature elevation of the entire brain for prolonged periods of time would result in adverse side effects and toxicities to patients. To provide a more targeted hyperthermic effect, MNPs may be delivered to brain tumors, followed by treatment with alternating magnetic fields (AMFs). Exposure of MNPs to AMFs results in the production of heat via Brownian Néel relaxation. This entire process of thermotherapy creates a local hyperthermia in brain tumors, while minimizing heating of the surrounding normal brain tissue (Fig. 7.10). Additionally, local or regional hyperthermia can result in elevated blood flow that might improve the delivery of chemotherapy to the brain and lead to a greater antitumor effect.

The distribution and density of MNPs can also be mapped following administration through CT imaging, in order to facilitate the planning of treatment parameters for the subsequent AMF exposure. For instance, with the density distribution of the nanoparticles, their known specific absorption rate, as well as the estimated perfusion within the tumor area, one group was able to use the bioheat transfer equation to determine the amount of heat generation within the target tissue as a function of magnetic field strength. This helped to establish an appropriate magnetic field strength to use for thermotherapy, such that a temperature of 43°C was not exceeded beyond a margin of 2 cm around the tumor [62]. This treatment planning process, referred to as postinstillation analysis, will likely play an important role in balancing the efficacy and safety of thermotherapy in brain tumor therapy.

GENE-BASED THERAPIES

Gene-based therapies encompass a variety of techniques that may be used modulate gene expression in specific cells to ultimately effectuate tumor regression. This can be accomplished by introducing exogenous nucleic acids such as DNA, small interfering RNA (siRNA), or antisense oligonucleotides (AONs) into cells. Given their large size, susceptibility to degradation, low transfection efficiency, and negative charge, these macromolecules require efficient delivery vectors to realize their true therapeutic potential [63].

Cationic polymers have been studied as nucleic acid delivery vehicles, as their positive charge facilitates the formation of nanocomplexes with negatively charged nucleic acids. At the same, these cationic polymers can associate with the negatively charged glycoproteins that are more highly expressed on brain tumor cells and tumor vasculature, compared to normal brain tissue. One study

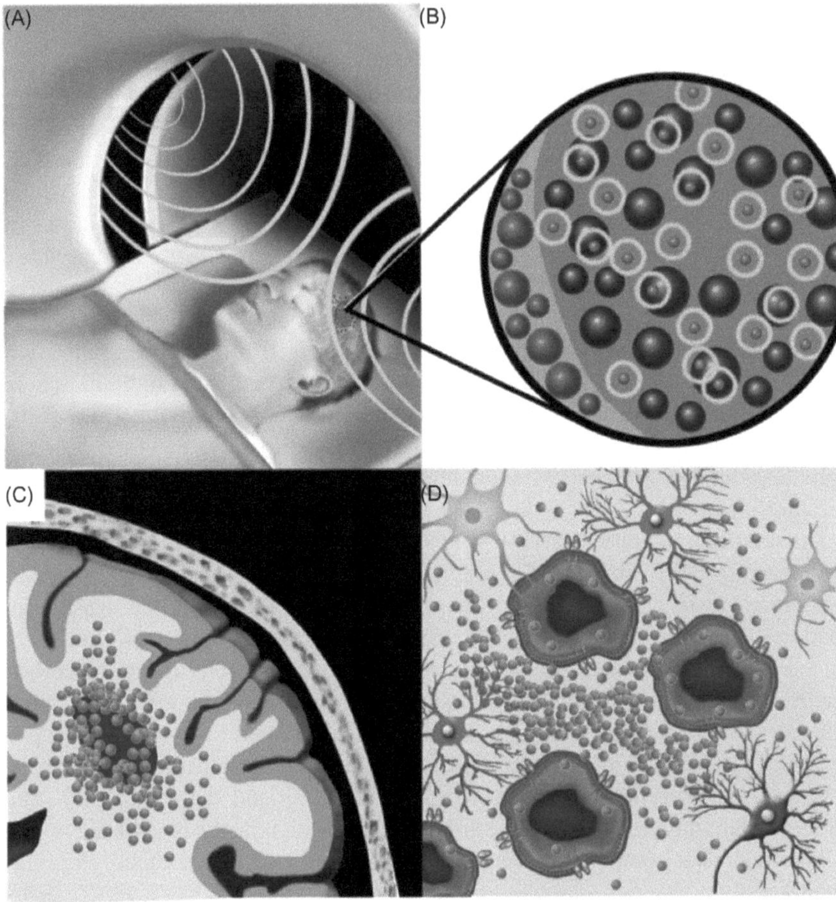

FIGURE 7.10

Local thermotherapy for a patient with a malignant brain tumor after MNP administration. (A) The patient is exposed to an alternating magnetic fields (*yellow* (white in print versions)). (B) Oscillation of the MNPs (*yellow* (white in print versions)) within the region of the tumor generates a local hyperthermia. (C) Local implantation of the MNPs within and adjacent to the brain tumor can provide a targeted hyperthermic effect. (D) Brain tumor cells (*purple* (dark gray in print versions)) may be more susceptible to the effects of local hyperthermia as a result of higher intracellular MNP uptake and sensitivity to temperature changes [61].

evaluated the efficacy of cationic albumin-conjugated PEGylated nanoparticles containing a plasmid encoding proapoptotic Apo2 ligand/tumor necrosis factor-related apoptosis-inducing ligand (Apo2L/TRAIL) in a mouse glioma model. At 30 minutes after intravenous injection, the nanoparticles were found to colocalize with glycoproteins in brain and tumor microvasculature and, via adsorptive-mediated transcytosis, accumulate in tumor cells. At 24 and 48 hours postinjection, hTRAIL mRNA and protein levels were significantly higher in tumors than in normal brain. At 14 days postinjection, repeated injections of CBSA-NP-hTRAIL had induced apoptosis and significantly delayed tumor growth in vivo [64].

Alternative approaches to gene-based therapy involve the use of nanoparticles to deliver AONs to cells. AONs bind to mRNA and prevent their translation into functional protein, effectively silencing a specific gene at the posttranscriptional level. For instance, polybutyl cyanoacrylate nanoparticles were used to deliver AONs against transforming growth factor-β (TGF-β) to brain tumors [65]. TGF-β secretion into the tumor microenvironment by tumor cells and certain immune cells is a major cause of localized immune suppression, which prevents immune-mediated destruction of tumors. TGF-β can inhibit T-cell and B-cell activation and proliferation, and suppresses natural killer cell activity impairment of tumor immune surveillance. In combination with a vaccine consisting of killed Newcastle-virus-infected tumor cells, the AON-PBCA formulation was able to decrease TGF-β plasma levels and increase activated CD25+ T lymphocytes that are important mediators of antitumor immunity. Interestingly the investigators found a positive correlation between the levels of CD25+ T lymphocytes and survival, which suggests that disinhibition of the intratumoral immune cells could significantly improve outcome in patients with brain tumors. This concept forms the basis of immunotherapy, in which the immune system activity is potentiated in an effort to eliminate tumors. Interestingly the immunotherapeutic approach to brain tumor therapy could be significantly bolstered by the use of nanoparticles that selectively target immune cells. For example, multi-walled carbon nanotubes (MWCNTs) appear to specifically target phagocytic cells within the brain [66]. One study showed that MWCNTs were phagocytosed predominantly by microglia, and resulted in a further influx of neutrophils, lymphocytes, and brain macrophages into the tumor site (Fig. 7.11) [67]. As microglia and other brain macrophage populations, such as perivascular macrophages, are partly responsible for the secretion of immunosuppressive cytokines into the tumor stroma [68,69], the ability to direct AONs and other therapeutic molecules to these cell populations with the use of MWCNTs could significantly facilitate brain tumor immunotherapy.

Gene-based therapy can also be targeted to BCICs, which are responsible for generating the bulk of the tumor. For instance, one group found that biodegradable cationic poly(β-amino ester)s (PBAEs) showed an inherent specificity for human primary BCICs. By systematically varying the structural components of the PBAEs—such as polymer backbone, side chain, polymer terminal group, and degradable linkage—they found that PBAEs self-assembled with plasmids encoding a DsRed fluorescent marker to form nanoparticles, with up to 100 plasmids encapsulated per nanoparticle. Transfection efficacy was demonstrated by DsRed transgene expression in tumor cells after intratumoral injection into an orthotopic murine model of human glioblastoma. Interestingly, the intrinsic specificity for human BCICs over normal human neural progenitor cells was not due to

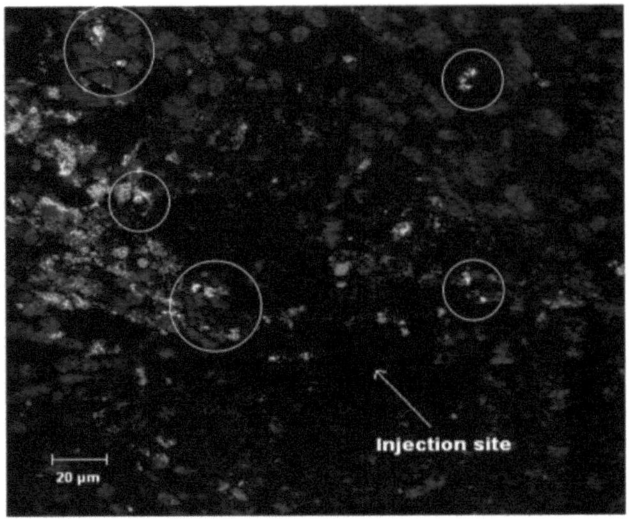

FIGURE 7.11

In vivo detection of MWCNT. MWCNTs are stained with PKH (*red* (dark gray in print versions)), CD68+ macrophages and microglia with FITC (*green* (white in print versions)), and tumor nuclei with DAPI (*blue* (light gray in print versions)). MWCNT-PKH-positive CD68+ cells are seen throughout the tumor and its periphery (*circles*), suggesting that macrophages and microglia efficiently phagocytose MWCNTs [67].

differences in cellular growth rate or total cellular uptake of nanoparticles. This was particularly evident at lower doses, where neural progenitor cells were transfected very poorly while significant transfection was observed for the BCIC samples. Moreover, the PBAE/DNA nanoparticles could be fabricated, lyophilized, and stored for at least 2 years without compromising their transfection capacity [70]. The specificity of transfection at low doses and with convenient storage properties are important practical considerations in the development of new medical technologies.

CHALLENGES

NEUROTOXICITY OF NANOPARTICLES

The biggest challenge in the application of nanotechnology to neurooncology is the potential neurotoxicity of nanoparticles. For instance, graphene is known to be cytotoxic. The hydrophobic interactions formed with membrane lipids, adsorption of biological molecules, and generation of ROS, which is unbalanced by cellular antioxidant enzymes, are among the various mechanisms of cytotoxicity mediated by graphene materials. In addition, graphene appears to have several genotoxic effects. Specifically, it is able to cause DNA fragmentation and chromosomal aberrations, even at low doses and short periods of exposure. The size, stiffness, hydrophobicity, surface functionalization, and dose of nanoparticles are major determinants of toxicity. Crucially the small size of nanoparticles means that a larger surface area and particle number are exposed to the body per unit mass compared to bulk agents. This increases the potential reactivity of nanoparticles, with subsequent implications on potential toxicity.

Moreover, uncertainties remain over the long-term effects of nanoparticles on the integrity of the BBB. Further study into the structural and functional consequences of nanoparticles on the BBB are necessary in order to identify potentially fatal side effects, such as brain tissue edema or unregulated entry of toxic molecules into the CNS. Ultimately the only common physiochemical property across the diverse nanomaterials that have been synthesized to date is a small size (1−100 nm). In addition, different nanomaterials interact with the microenvironment in unique ways, such that changes to size, shape, or aggregation state vary widely. As such, the safety of each material, along with slight variations in structure that might have dramatic and unpredictable effects on toxicity, must be separately and thoroughly assessed. This would enable a thorough clinical evaluation of the risks and benefits of different nanoparticle formulations. However, attention must be given to the possibility that nanoparticles might have very differently in different organisms, such that safety in animal studies does not necessarily guarantee safety in humans.

Moreover, the use of nanoparticles, particularly when systemically administered, poses the risk of a widespread immune response. For instance, systemically administered CNTs have been shown to activate both the classical and alternative arms of the human complement cascade. High doses of nanoparticles, which may be required for maximal therapeutic efficacy, may thus cause inflammatory pathology and widespread granuloma formation [22]. Some patients may also show adverse allergic reactions to certain nanomaterials. For instance, carbon based nanomaterials generate a Th2 immune response mediated through the IL33/ST2 axis, and this is in turn associated with the development of immediate hypersensitivity reactions [71]. At the same time, it is important to recognize that nanoparticles are in fact surrounded by a protein corona in vivo. It is this protein corona, rather than the nanomaterial itself, that determines the immune response to the nanoparticle.

BIOAVAILABILITY

The nanoparticle-directed immune response could be particularly problematic if it results in the premature clearance of nanoparticles before they can mediate their therapeutic effects. Moreover, successive therapies with the same nanoparticles would result in progressively faster immune clearance and lower therapeutic efficacy. In order to enhance the circulating time of nanoparticles and increase their delivery to target tissue, several groups have studied the ability of PEG coatings to reduce protein adsorption, diminish nonspecific interactions, and improve pharmacokinetics. PEG-coated nanoparticles prepared from a poly(PEG cyanoacrylatecohexadecyl cyanoacrylate) copolymer show not only greater accumulation in the brain compared to non-PEG-coated nanoparticles but also longer circulated times in the blood stream and reduced hepatic sequestration. One study found that a methoxy-PEG-alkyl-thiol coating could reduce macrophage uptake of GNPs, as the intervening alkyl between the PEG and thiol moieties forms an effective hydrophobic shield that prevents protein adsorption [72].

Apart from avoiding immune clearance by the liver and macrophages, which form part of the reticuloendothelial system of the body, bioavailability depends on their ability to access their target tissue. While brain specificity can be an engineered inherent property of nanoparticles, their bioavailability in the brain could be further enhanced with the development of clinical interventions to manipulate the BBB prior to nanoparticle delivery.

FUTURE DIRECTIONS

It is important to remember that brain tumors are a heterogeneous group of neoplasms. The tumor grade, patient fitness, and anatomical location of tumor significantly affect the kind of treatment that should be instituted, as well as the likelihood of treatment success. For instance, most studies of PDT have been restricted to treating supratentorial neoplasms that arise above the tentorium cerebelli, away from functional cortex and the brainstem. Infratentorial malignancies in the posterior fossa are in close proximity to the brainstem, such that therapies such as PDT may cause immediate or long-term complications [73]. One must thus avoid making the assumption that therapies that apply for one type of brain tumor necessarily apply for other tumor types or for the same tumor type in a different anatomical location.

Given the complexity of the technology and medical concepts involved, clinical trials in nanomedicine will raise unique challenges with regard to risk minimization and management, as well as risk communication with human subjects and the wider public. Rigorous Phase IV postmarketing studies will be necessary to understand the long-term safety and efficacy of nanomedical products. Future work should also look into the possibility of identifying biomarkers to determine which patients are likely to benefit from different treatments, and which are likely to be adversely affected. At the crux of these multiple considerations is the need to continually improve our understanding of the abiotic:biotic interface. Armed with this knowledge and an ever-increasing range of nanotechnologies, we will find ourselves in a better position to help patients in the battle against brain cancer.

REFERENCES

[1] Jemal A, Center MM, DeSantis C, Ward EM. Global patterns of cancer incidence and mortality rates and trends. Cancer Epidemiol Biomarkers Prev 2010;19(8):1893–907.

[2] Gavrilovic IT, Posner JB. Brain metastases: epidemiology and pathophysiology. J Neurooncol 2005;75(1):5–14.

[3] Brandes AA. Introduction: future trends in the treatment of brain tumors. Semin Oncol 2003;30:1–3.

[4] Harris RE. Global epidemiology of cancer. Burlington, MA: Jones & Bartlett Learning; 2016; xiii, 429 pages.

[5] Sathornsumetee S, Rich JN. New approaches to primary brain tumor treatment. Anticancer Drugs 2006;17(9):1003–16.

[6] Woodworth GF, Dunn GP, Nance EA, Hanes J, Brem H. Emerging insights into barriers to effective brain tumor therapeutics. Front Oncol 2014;4:126.

[7] Mattei TA, Rehman AA. "Extremely minimally invasive": recent advances in nanotechnology research and future applications in neurosurgery. Neurosurg Rev 2015;38(1):27–37 [discussion].

[8] de Boer AG, Gaillard PJ. Drug targeting to the brain. Ann Rev Pharmacol Toxicol 2007;47:323–55.

[9] Agarwala SS, Kirkwood JM. Temozolomide, a novel alkylating agent with activity in the central nervous system, may improve the treatment of advanced metastatic melanoma. Oncologist 2000;5(2):144–51.

[10] Huang G, Zhang N, Bi X, Dou M. Solid lipid nanoparticles of temozolomide: potential reduction of cardial and nephric toxicity. Int J Pharm 2008;355(1–2):314–20.

[11] Wei X, Chen X, Ying M, Lu W. Brain tumor-targeted drug delivery strategies. Acta Pharm Sin B 2014;4(3):193–201.

[12] Lu W, Wan J, Zhang Q, She Z, Jiang X. Aclarubicin-loaded cationic albumin-conjugated pegylated nanoparticle for glioma chemotherapy in rats. Int J Cancer 2007;120(2):420–31.

[13] Qin Y, Fan W, Chen H, Yao N, Tang W, Tang J, et al. In vitro and in vivo investigation of glucose-mediated brain-targeting liposomes. J Drug Target 2010;18(7): 536–49.

[14] Michaelis K, Hoffmann MM, Dreis S, Herbert E, Alyautdin RN, Michaelis M, et al. Covalent linkage of apolipoprotein e to albumin nanoparticles strongly enhances drug transport into the brain. J Pharmacol Exp Ther 2006;317(3):1246–53.

[15] Kreuter J, Shamenkov D, Petrov V, Ramge P, Cychutek K, Koch-Brandt C, et al. Apolipoprotein-mediated transport of nanoparticle-bound drugs across the blood-brain barrier. J Drug Target 2002;10(4):317–25.

[16] Tzeng SY, Green JJ. Therapeutic nanomedicine for brain cancer. Ther Deliv 2013;4(6):687–704.

[17] Singh SK, Hawkins C, Clarke ID, Squire JA, Bayani J, Hide T, et al. Identification of human brain tumour initiating cells. Nature 2004;432(7015):396–401.

[18] Beck S, Jin X, Yin J, Kim SH, Lee NK, Oh SY, et al. Identification of a peptide that interacts with Nestin protein expressed in brain cancer stem cells. Biomaterials 2011;32(33):8518–28.

[19] Goffart N, Kroonen J, Rogister B. Glioblastoma-initiating cells: relationship with neural stem cells and the micro-environment. Cancers (Basel) 2013;5(3):1049–71.

[20] Richards Grayson AC, Choi IS, Tyler BM, Wang PP, Brem H, Cima MJ, et al. Multi-pulse drug delivery from a resorbable polymeric microchip device. Nat Mater 2003;2(11):767–72.

[21] Kim GY, Tyler BM, Tupper MM, Karp JM, Langer RS, Brem H, et al. Resorbable polymer microchips releasing BCNU inhibit tumor growth in the rat 9L flank model. J Control Release 2007;123(2):172–8.

[22] Liang XJ, Chen C, Zhao Y, Wang PC. Circumventing tumor resistance to chemotherapy by nanotechnology. Methods Mol Biol 2010;596:467–88.

[23] Koziara JM, Lockman PR, Allen DD, Mumper RJ. Paclitaxel nanoparticles for the potential treatment of brain tumors. J Control Release 2004;99(2):259–69.

[24] Qian L, Zheng J, Wang K, Tang Y, Zhang X, Zhang H, et al. Cationic core-shell nanoparticles with carmustine contained within *O*(6)-benzylguanine shell for glioma therapy. Biomaterials 2013;34(35):8968–78.

[25] Hegi ME, Diserens AC, Gorlia T, Hamou MF, de Tribolet N, Weller M, et al. MGMT gene silencing and benefit from temozolomide in glioblastoma. N Engl J Med 2005;352(10):997–1003.

[26] Weller M, Stupp R, Reifenberger G, Brandes AA, van den Bent MJ, Wick W, et al. MGMT promoter methylation in malignant gliomas: ready for personalized medicine? Nat Rev Neurol 2010;6(1):39–51.

[27] Thon N, Kreth S, Kreth FW. Personalized treatment strategies in glioblastoma: MGMT promoter methylation status. Onco Targets Ther 2013;6:1363–72.

[28] Donson AM, Addo-Yobo SO, Handler MH, Gore L, Foreman NK. MGMT promoter methylation correlates with survival benefit and sensitivity to temozolomide in pediatric glioblastoma. Pediatr Blood Cancer 2007;48(4):403–7.

[29] Huang FY, Lee TW, Chang CH, Chen LC, Hsu WH, Chang CW, et al. Evaluation of 188Re-labeled PEGylated nanoliposome as a radionuclide therapeutic agent in an orthotopic glioma-bearing rat model. Int J Nanomed 2015;10:463–73.

[30] Kelkar SS, Reineke TM. Theranostics: combining imaging and therapy. Bioconjug Chem 2011;22(10):1879–903.

[31] Su XY, Liu PD, Wu H, Gu N. Enhancement of radiosensitization by metal-based nanoparticles in cancer radiation therapy. Cancer Biol Med 2014;11(2):86–91.

[32] Joh DY, Sun L, Stangl M, Al Zaki A, Murty S, Santoiemma PP, et al. Selective targeting of brain tumors with gold nanoparticle-induced radiosensitization. PLoS One 2013;8(4):e62425.

[33] Dorsey JF, Sun L, Joh DY, Witztum A, Kao GD, Alonso-Basanta M, et al. Gold nanoparticles in radiation research: potential applications for imaging and radiosensitization. Transl Cancer Res 2013;2(4):280–91.

[34] Santos T, Ferreira R, Maia J, Agasse F, Xapelli S, Cortes L, et al. Polymeric nanoparticles to control the differentiation of neural stem cells in the subventricular zone of the brain. ACS Nano 2012;6(12):10463–74.

[35] Delcroix GJ, Jacquart M, Lemaire L, Sindji L, Franconi F, Le Jeune JJ, et al. Mesenchymal and neural stem cells labeled with HEDP-coated SPIO nanoparticles: in vitro characterization and migration potential in rat brain. Brain Res 2009; 1255:18–31.

[36] Namgung S, Baik KY, Park J, Hong S. Controlling the growth and differentiation of human mesenchymal stem cells by the arrangement of individual carbon nanotubes. ACS Nano 2011;5(9):7383–90.

[37] Fazekas JT. Treatment of grades I and II brain astrocytomas. The role of radiotherapy. Int J Radiat Oncol Biol Phys 1977;2(7–8):661–6.

[38] Leibel SA, Sheline GE, Wara WM, Boldrey EB, Nielsen SL. The role of radiation therapy in the treatment of astrocytomas. Cancer 1975;35(6):1551–7.
[39] Moore GE, Peyton WT, et al. The clinical use of fluorescein in neurosurgery: the localization of brain tumors. J Neurosurg 1948;5(4):392–8.
[40] Orringer DA, Koo YE, Chen T, Kopelman R, Sagher O, Philbert MA. Small solutions for big problems: the application of nanoparticles to brain tumor diagnosis and therapy. Clin Pharmacol Ther 2009;85(5):531–4.
[41] Stummer W, Tonn JC, Goetz C, Ullrich W, Stepp H, Bink A, et al. 5-Aminolevulinic acid-derived tumor fluorescence: the diagnostic accuracy of visible fluorescence qualities as corroborated by spectrometry and histology and postoperative imaging. Neurosurgery 2014;74(3):310–19 [discussion 9–20].
[42] Karabeber H, Huang R, Iacono P, Samii JM, Pitter K, Holland EC, et al. Guiding brain tumor resection using surface-enhanced Raman scattering nanoparticles and a hand-held Raman scanner. ACS Nano 2014;8(10):9755–66.
[43] Arlett JL, Myers EB, Roukes ML. Comparative advantages of mechanical biosensors. Nat Nanotechnol 2011;6(4):203–15.
[44] Kang SK, Murphy RK, Hwang SW, Lee SM, Harburg DV, Krueger NA, et al. Bioresorbable silicon electronic sensors for the brain. Nature 2016;530(7588):71–6.
[45] Clausen I, Glott T. Development of clinically relevant implantable pressure sensors: perspectives and challenges. Sensors (Basel) 2014;14(9):17686–702.
[46] Chen X, Kis A, Zettl A, Bertozzi CR. A cell nanoinjector based on carbon nanotubes. Proc Natl Acad Sci USA 2007;104(20):8218–22.
[47] Kumar AJ, Leeds NE, Fuller GN, Van Tassel P, Maor MH, Sawaya RE, et al. Malignant gliomas: MR imaging spectrum of radiation therapy- and chemotherapy-induced necrosis of the brain after treatment. Radiology 2000;217(2):377–84.
[48] Mullins ME, Barest GD, Schaefer PW, Hochberg FH, Gonzalez RG, Lev MH. Radiation necrosis versus glioma recurrence: conventional MR imaging clues to diagnosis. AJNR Am J Neuroradiol 2005;26(8):1967–72.
[49] Gahramanov S, Raslan AM, Muldoon LL, Hamilton BE, Rooney WD, Varallyay CG, et al. Potential for differentiation of pseudoprogression from true tumor progression with dynamic susceptibility-weighted contrast-enhanced magnetic resonance imaging using ferumoxytol vs. gadoteridol: a pilot study. Int J Radiat Oncol Biol Phys 2011;79(2):514–23.
[50] Chan WC, Nie S. Quantum dot bioconjugates for ultrasensitive nonisotopic detection. Science 1998;281(5385):2016–18.
[51] Murphy CJ. Optical sensing with quantum dots. Anal Chem 2002;74(19):520A–526AA.
[52] Vanmaekelbergh D, Liljeroth P. Electron-conducting quantum dot solids: novel materials based on colloidal semiconductor nanocrystals. Chem Soc Rev 2005;34(4):299–312.
[53] West JL, Halas NJ. Engineered nanomaterials for biophotonics applications: improving sensing, imaging, and therapeutics. Annu Rev Biomed Eng 2003;5:285–92.
[54] Walling MA, Novak JA, Shepard JRE. Quantum dots for live cell and in vivo imaging. Int J Mol Sci 2009;10(2):441–91.
[55] Bonneau S, Dahan M, Cohen LD. Single quantum dot tracking based on perceptual grouping using minimal paths in a spatiotemporal volume. IEEE Trans Image Process 2005;14(9):1384–95.
[56] Bonneau S, Cohen L, Dahan M, editors. A multiple target approach for single quantum dot tracking. In: IEEE international symposium on biomedical imaging: nano to macro, April 2004; 2004

[57] Saxton MJ, Jacobson K. Single-particle tracking: applications to membrane dynamics. Annu Rev Biophys Biomol Struct 1997;26:373–99.
[58] Ballou B, Lagerholm BC, Ernst LA, Bruchez MP, Waggoner AS. Noninvasive imaging of quantum dots in mice. Bioconjug Chem 2004;15(1):79–86.
[59] Tian B, Wang C, Zhang S, Feng L, Liu Z. Photothermally enhanced photodynamic therapy delivered by nano-graphene oxide. ACS Nano 2011;5(9):7000–9.
[60] Akhavan O, Ghaderi E. Graphene nanomesh promises extremely efficient in vivo photothermal therapy. Small 2013;9(21):3593–601.
[61] Mahmoudi K, Hadjipanayis CG. The application of magnetic nanoparticles for the treatment of brain tumors. Front Chem 2014;2:109.
[62] Maier-Hauff K, Ulrich F, Nestler D, Niehoff H, Wust P, Thiesen B, et al. Efficacy and safety of intratumoral thermotherapy using magnetic iron-oxide nanoparticles combined with external beam radiotherapy on patients with recurrent glioblastoma multiforme. J Neurooncol 2011;103(2):317–24.
[63] Yin H, Kanasty RL, Eltoukhy AA, Vegas AJ, Dorkin JR, Anderson DG. Non-viral vectors for gene-based therapy. Nat Rev Genet 2014;15(8):541–55.
[64] Lu W, Sun Q, Wan J, She Z, Jiang XG. Cationic albumin-conjugated pegylated nanoparticles allow gene delivery into brain tumors via intravenous administration. Cancer Res 2006;66(24):11878–87.
[65] Schneider T, Becker A, Ringe K, Reinhold A, Firsching R, Sabel BA. Brain tumor therapy by combined vaccination and antisense oligonucleotide delivery with nanoparticles. J Neuroimmunol 2008;195(1–2):21–7.
[66] Kateb B, Van Handel M, Zhang L, Bronikowski MJ, Manohara H, Badie B. Internalization of MWCNTs by microglia: possible application in immunotherapy of brain tumors. Neuroimage 2007;37(Suppl. 1):S9–17.
[67] VanHandel M, Alizadeh D, Zhang L, Kateb B, Bronikowski M, Manohara H, et al. Selective uptake of multi-walled carbon nanotubes by tumor macrophages in a murine glioma model. J Neuroimmunol 2009;208(1–2):3–9.
[68] Wu A, Wei J, Kong LY, Wang Y, Priebe W, Qiao W, et al. Glioma cancer stem cells induce immunosuppressive macrophages/microglia. Neuro Oncol 2010;12(11): 1113–25.
[69] Zhang L, Handel MV, Schartner JM, Hagar A, Allen G, Curet M, et al. Regulation of IL-10 expression by upstream stimulating factor (USF-1) in glioma-associated microglia. J Neuroimmunol 2007;184(1–2):188–97.
[70] Guerrero-Cazares H, Tzeng SY, Young NP, Abutaleb AO, Quinones-Hinojosa A, Green JJ. Biodegradable polymeric nanoparticles show high efficacy and specificity at DNA delivery to human glioblastoma in vitro and in vivo. ACS Nano 2014;8(5): 5141–53.
[71] Caffo M, Merlo L, Marino D, Caruso G. Graphene in neurosurgery: the beginning of a new era. Nanomedicine (Lond) 2015;10(4):615–25.
[72] Larson TA, Joshi PP, Sokolov K. Preventing protein adsorption and macrophage uptake of gold nanoparticles via a hydrophobic shield. ACS Nano 2012;6(10): 9182–90.
[73] Quirk BJ, Brandal G, Donlon S, Vera JC, Mang TS, Foy AB, et al. Photodynamic therapy (PDT) for malignant brain tumors—where do we stand? Photodiagn Photodyn Ther 2015;12(3):530–44.

CHAPTER

8 Nanotechnologies for brain tumor therapy

G. Caruso, L. Merlo and M. Caffo

University of Messina, Messina, Italy

CHAPTER OUTLINE

OVERVIEW ON BRAIN GLIOMAS

Although cerebral gliomas represent only 2% of cancers, they account for about 45% of all primary central nervous system (CNS) tumors [1] being among the most deadly neoplasms.

In big series of patients harboring a malignant brain tumor, survival rate goes between 2% and 4% at 3–5 years from treatment [2,3]. This percentage goes up (until 11% of survival at 3 years) when considering smaller series [4].

Nowadays the great majority of patients harboring a malignant brain tumor receives the "Stupp protocol" (radiation + low-dose temozolomide (TMZ), followed by monthly TMZ) as initial treatment [5]. Survival data of 254 patients receiving this protocol showed that 16% of them were alive after 3 years, while only 9.8% after 5 years [6]. Although the 5-year survival rate was approximately double that of previous series, the survival curve showed no sign of plateau, which suggests that the asymptotic level of survival will eventually reach zero.

A different perspective has been provided by analyses examining the conditional probabilities of surviving in terms of additional years given different lengths of prior survival time. The largest conditional probability analysis comes from the American SEER database [7], which includes over 10,000 patients with glioblastoma (GBM) diagnosis from the years 1998 to 2008. The probabilities to survive an additional year were 53% from the time of diagnosis, 38% after 1 year of prior survival, 55% after 2 years, 70% after 3 years, 82% after 4 years, and 78% after 5 years. After 5 years of survival, the probability of surviving an

Nanotechnology in Cancer.

additional 5 years was 45%. The percentage of patients surviving 5 years was 6.2, which implies that the percentage surviving 10 years was approximately 3. In this kind of analysis, variables predicting survival, such as tumor grade and patient age, lost their prognostic value after 3–4 years of prior survival.

The implication of these analyses is that a small percentage of patients survives at least 5–10 years, but there continues to be a substantial rate of death even after extended survival, although the death rate substantially decreases the longer the prior survival.

It is also important to recognize that some of the late "relapses" may not be recurrences of the original tumor but new disease induced by radiation treatment. Experimental work with animal models supports the reality of this risk [8].

Long-term survival can also be due to particular characteristics of patients themselves or of treatments they received. The great majority of long-term survivors (LTS) have had systemic chemotherapy, although this may simply reflect the fact that chemotherapy has been part of the standard treatment. In fact, there is one report of three LTS (11, 16, and 18 years) of 71 patients receiving only brachytherapy [9].

Patient characteristics most common among LTS are young age at the time of diagnosis, higher Karnofsky performance status (KPS), methylation of the MGMT promoter gene, and a complete surgical resection. However, there are LTS who are exceptions to each of these generalizations. The most extensive report of the characteristics of LTS comes from the German Glioma Network, which compared 69 patients who survived more than 36 months with 257 patients who survived less than 36 months [10]. LTS patients were younger but not significantly different with respect to KPS. They were also only marginally more likely to have complete resection at the time of initial surgery, but were significantly more likely to have two or more surgical interventions. In addition, LTS patients were significantly more likely to have methylation of MGMT promoter gene, but less likely to have EGFR amplification. The difference in p53 mutations was not significant. The most consistent variable associated with LTS was isocitrate dehydrogenase (IDH) mutations, as 33% of LTS patients had the mutation, compared to only 4% in control patients. Presence of MGMT methylation and IDH mutations were highly correlated but still partially dissociable. Specifically, patients with IDH mutations but without MGMT methylation had the same prognosis as patients without IDH mutations. It is also noteworthy that there was a significant number of LTS who had no IDH mutations.

As of now, we have only a minimal basis for predicting which patients will receive a significant benefit of treatment, and even less basis for choosing treatments that will be most successful.

The biology of gliomas can help for a targeted therapy.

Gliomas can originate from neural stem cells, progenitor cells, or from de-differentiated mature neural cells transformed into cancer stem cells [11]. Currently the most widely used gliomas histological and molecular classification, and grading system is that of the World Health Organization (WHO) [12] where astrocytomas are the most common subtype. Overall, GBM has the worst prognosis. According to the new molecular WHO classification, there are GBM IDH-wildtype and GBM IDH-mutant.

On the base of the course of the tumor, it is possible to distinguish other two subtypes of GBM: primary and secondary.

Primary GBM develops rapidly de novo in elderly patients, without clinical or histologic evidence of a less malignant precursor lesion and represents the vast majority of GBMs (~90%). Secondary GBMs progress from previous low-grade diffuse astrocytoma or anaplastic astrocytoma and manifest in younger patients, have a lesser degree of necrosis, are preferentially located in the frontal lobe, and carry a better prognosis. Histologically, primary and secondary GBMs are similar. They differ in their genetic and epigenetic profiles. Decisive genetic signposts of secondary GBMs are IDH1 mutations, which are absent in the primary and which are associated with a hypermethylation phenotype. IDH1 mutations are the earliest detectable genetic alterations in precursor low-grade diffuse astrocytomas and in oligodendrogliomas, indicating that these tumors derive from neural precursor cells that differ from those of primary GBMs. IDH1 mutations are diagnostic molecular markers of secondary GBMs and more reliable and objective than clinical criteria. Thus despite a similar histologic appearance, primary and secondary GBMs are distinct tumor entities that originate from different precursor cells and may require different therapeutic approaches [13].

Conventional brain tumor treatments include surgery, radiation therapy, and chemotherapy [14–16]. Each of them has its own disadvantages. Surgical treatment usually represents the first approach: it is invasive but warrants a histological diagnosis. Radiation and chemotherapy are less invasive and can be used alone or often as adjuvant therapy to obtain a prolonged progression-free survival. Radiotherapy gives a better control of seizures with no substantial differences in overall survival [17] but higher risk to develop post-radiation leukoencephalopathy [18]. The effectiveness of systemic chemotherapy is limited by toxic effects on healthy cells, generally resulting in morbidity or mortality of the patient. Moreover the presence of the blood–brain barrier (BBB) limits the passage of a wide variety of anticancer agents [19]. Several drugs possess poor solubility, high toxicity and high dosage, nonspecific delivery, in vivo degradation, and short circulating half-lives.

Despite intensive research for more effective treatments, aggressive multimodal protocols have extended the median survival but often with a significant impairment in the quality of life.

These disappointing results might be due to tumor invasion into functional brain tissue, lack of chemosensitivity, and shortcomings of the systemic delivery [20]. Some forms of systemic chemotherapy partially able to cross the BBB (nitrosoureas such as BCNU and lomustine, and alkylating agents such as TMZ) provide only modest advantages when added to radiotherapy [21]. Moreover, low-molecular-weight chemotherapeutics do not achieve and maintain effective steady-state concentrations within GBM cells because of short blood half-lives [22]. Currently, most therapeutic agents targeting brain tumors are delivered systemically by intravenous and oral route that have limitations reducing the effectiveness of these drugs. The limited success in treating brain tumors results from the tumor cell chemoresistance (natural or acquired), poor selectivity of the antitumor drugs, and presence of the BBB. Multidrug resistance (MDR) is ascribed to ineffective drug delivery to tumor tissues and tumor cells. The advance of the drug delivery to the tumor sites and the decrease of the MDR-based drug efflux can represent ideal solution [23].

Devices that are able to deliver efficacious drugs' concentrations into the brain tumor cells are nanoparticles-derived.

Nanotechnology is becoming always more diffuse and used in lot of fields: from electronics to medicine. Nanodevices are sometimes very expensive.

Analyzing the cost of treatment of a brain tumor patient, the items to consider are bed days, investigations, surgery, radiotherapy, chemotherapy, and outpatient follow-up. In a British database the mean costs for each of these components were 442 Pounds for neuroradiological investigations, 2,407 Pounds for neurosurgical bed days, 2,068 Pounds for neurosurgery, 434 Pounds for neuropathology, 8,832 Pounds for radiotherapy, 440 Pounds for chemotherapy, and 1,078 Pounds for outpatients follow-up. Considering the total treatment costs per patient, the range goes from 1,978 to 26,980 Pounds [24].

Large-scale manufacturing remains a costly and challenging aspect in the clinical translation of biotargeted nanomedicines. At present, there are few cost–benefit studies available for nanomedicine products. As an example the cost-effectiveness of Doxil (Janssen Biotech Inc., PA, USA; PEGylated liposomal doxorubicin) and Abraxane (Celgene Corporation, NJ, USA; nano-albumin bound paclitaxel) compared with their conventional standard-of-care generic alternatives, doxorubicin, and paclitaxel, respectively. In 2009 the average cost per dose of Doxil was US$5,594 compared with $62–162 for doxorubicin, and the average cost per dose for Abraxane was $5,054 compared with $90–454 for paclitaxel. Notable health- and cost-related benefits of Doxil and Abraxane are lower cardiac toxicity and reduced vehicle toxicity, respectively. Although neither nanomedicine products have shown an increase in overall patient survival, the reduction in toxicities and their associated cost have largely justified the higher cost [25]. Increasing pressure to reduce healthcare costs puts an even greater burden on the nanomedicine innovator to justify the real cost-to-benefit ratio.

The manufacturing also changes the price of nanomaterials. For example, graphene can be synthesized either through chemical vapor deposition or from the liquid-phase exfoliation of graphite, the latter being the most widely adopted procedure because technologically simple, highly efficient, and low-cost [26]. Such optimized technique of synthesis caused its overall costs dramatically to drop: while in 2008 graphene was one of the most expensive materials, costing about $100 million per cm^2, now reduced graphene oxide costs approximately $270 per gram [27].

In order to lower the costs, many of the companies that are currently developing biotargeted cancer therapies are smaller start-ups featuring pipelines based on technologies originally developed in academic laboratories, rendering high development costs even more daunting. There is a movement toward collaborative efforts between large and small pharma, government, nonprofit agencies, and venture capital firms to defray costs of new therapeutic development [25].

BLOOD–BRAIN BARRIER

The brain is a unique organ highly protected by two major barriers, the largest BBB and the blood–cerebrospinal fluid barrier. Overcoming the BBB would improve significantly the delivery of targeted drugs to malignant glioma cells.

The BBB is responsible for maintenance of neuronal microenvironment, tissue homeostasis, vasotonous regulation, fibrinolysis and coagulation, and blood cell activation and migration during physiological and pathological processes. Physiologically, BBB is composed of an inner endothelial cell layer forming the wall of the capillary and contains tight junctions (TJs) and a basement membrane upon which pericytes and astrocytic feet processes lie [28]. The BBB endothelial cells differ from endothelial cells in the rest of the body by the absence of fenestrations, more extensive TJs, and sparse pinocytic vesicular transport. Endothelial cells TJs limit the paracellular flux of hydrophilic molecules across the barrier (Fig. 8.1).

The TJs are located on the apical region of endothelial cells and are formed by a complex network made of a series of parallel, interconnected, transmembrane,

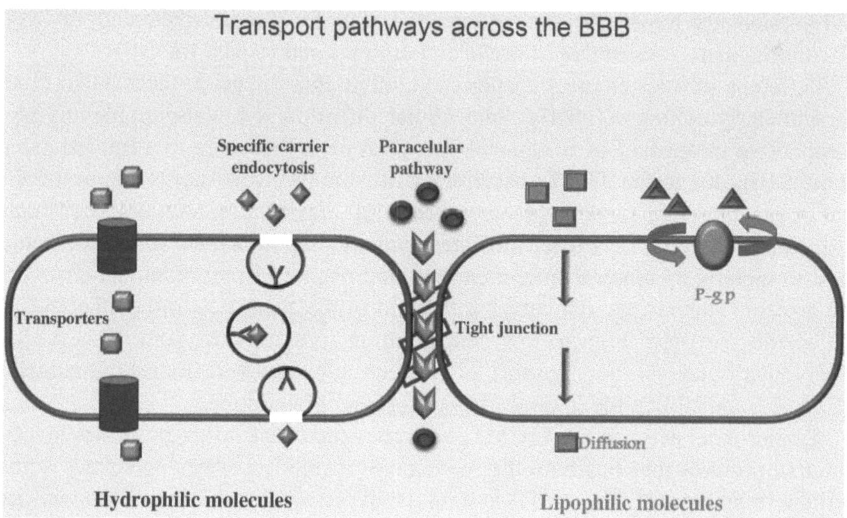

FIGURE 8.1

Molecular transport across the blood–brain barrier.

and cytoplasmatic strands of proteins [29]. They consist of three integral membrane proteins, namely, claudin, occludin, and junction adhesion molecules, and a number of cytoplasmic accessory proteins. The tightness of the BBB is due to the physical complexity of its junctional structure and the molecular substructure, in particular, the presence of transmembrane proteins, which help to seal the intercellular cleft. Adherens junctions (AJs) located below the TJs in the basal region of the lateral plasmamembrane are composed of transmembrane glycoproteins (cadherins) linked to the cytoskeleton by cytoplasmatic proteins, thus providing an additional tightening structure between the adjacent endothelial cells at the BBB [30].

Despite the rapid development in understanding the molecular structure and receptor expressed of the BBB, many of the CNS-associated diseases remain undertreated by effective therapies. The majority of drugs and large-molecular-weight particulate agents such as recombinant proteins, peptides, monoclonal antibodies, small-interfering RNA (siRNA), and gene therapeutics do not readily permeate into the brain parenchyma.

There are different mechanisms by which molecules can cross the BBB. Passive diffusion is dependent on concentration gradient, lipophilicity, and molecular weight that allows molecules to move between cells (paracellular way) or across cells (transcellular way) down their electrochemical gradient without the requirement of metabolic energy. For example, small water-soluble molecules simply diffuse through the TJs. Small lipid-soluble substances like alcohol and steroid hormones penetrate transcellularly by dissolving in their lipid plasmamembrane. However the majority of small-molecule drugs have a higher molecular weight or current water solubility preventing their simple diffusion across the barrier. For almost all other substances, including essential materials such as glucose and aminoacids, transport proteins (carriers), specific receptor-mediated, or vesicular mechanisms (adsorptive transcytosis) are required to pass the BBB.

Different substances are transported through free diffusion mechanism either paracellularly or transcellularly. Paracellular diffusion is a nonsaturable and noncompetitive movement of compounds between cells. It occurs to a limited extent at the BBB, due to the TJs. Transcellular diffusion (transcytosis) is a nonsaturable and noncompetitive movement across cells of lipophilic substances. Facilitated diffusion is a form of carrier-mediated endocytosis in which solute molecules bind to specific membrane protein carriers that trigger a conformational change in the protein. This results in a carrying through of the substance to the other side of the membrane, from high to low concentration (passive diffusion). This mechanism contributes to the transport of various substances including aminoacids, nucleoside, small peptide, monocarboxylates, and glutathione.

Carrier-mediated transport (CMT) or carrier-mediated influx processes involve putative proteins that facilitate the movement of poorly permeable solutes across cellular membranes. The CMT system is expressed on both the luminal and abluminal membranes of the brain capillary endothelium and operates in both directions. CMT systems can be exploited for brain drug-delivery after reformulating the drug in such a way that the drug assumes a molecular structure mimicking that of the endogenous ligand. If compounds need to be moved against a concentration gradient, ATP may provide the energy to facilitate the process.

The uptake of nutrients from blood into the brain is facilitated by the solute carrier transporter families. These influx carriers are involved in the transport of a broad range of substrates including glucose, amino acids, nucleosides, fatty acids, minerals, and vitamins in various human tissues, including the brain.

The active efflux transport is responsible for extruding drugs from the brain and this mechanism is a major obstacle for the accumulation of a wide range of biologically active molecules in the brain. The ATP-binding cassette (ABC) transporter P-glycoprotein and multidrug-resistant protein represent the principle efflux mechanism of these agents [31]. The most abundantly present component of this system is efflux P-glycoprotein, which is a product of the ABCB1 gene. Inhibition of P-glycoprotein in preclinical studies has enhanced the penetration of paclitaxel into the brain, indicating the feasibility of achieving improved drug delivery to the brain by suppression of P-glycoprotein [32].

Endocytosis and transcytosis allow the internalization, sorting and trafficking of many plasma macromolecules. Endocytosis is a process where molecules from the circulation are internalized in vesicles and are directed to endosomes or lysosomes within the cell. Endocytosis can be isolated into bulk-phase (fluid phase or pinocytosis) endocytosis and mediated endocytosis (receptor and absorptive mediated). Bulk-phase endocytosis is the noncompetitive, nonsaturable, temperature- and energy-dependent nonspecific uptake of extracellular fluids. Transcytosis refers to the transcellular movement of molecules.

Receptor-mediated endocytosis or clathrin-dependent endocytosis provides for a highly specific and energy-mediated transport enabling eukaryotic cells to selective uptake macromolecules as specific cargo. Cells have different receptors for the uptake of many different types of ligands, including hormones, growth factors, enzymes, and plasma proteins. This process occurs at the brain for macromolecular substances, such as transferrin, insulin, leptin, IGF-I, and IGF-II, and is a highly specific type of energy-dependent transport [33].

Adsorptive endocytosis/transcytosis facilitates the transport of large peptides such as IgG, histone, albumin, native ferritin, horseradish peroxidase, and dextran. Adsorptive-mediated endocytosis is characterized by an electrostatic interaction between a positively charged substance and the negatively charged sites on the brain endothelial cell surface (e.g., glycoprotein) [34]. Adsorptive processes largely depend upon electrostatic interactions that allow the positively charged moiety of the substrate to bind to the negatively charged cell membrane. Receptor-mediated transport is mainly employed in the transport of macromolecules like peptides and proteins across the BBB, by conjugating the substance with ligands such as lactoferrin, transferrin, and insulin. It is an important transport mechanism of predominant interest in drug delivery.

Cell-mediated transcytosis is a recently identified route of drug transport across the BBB [35]. This transport route relies on immune cells such as monocytes or macrophages to cross the intact BBB. Unlike the aforementioned transport pathways, which normally permit only solute molecules with specific properties, cell-mediated transcytosis is unique in that it can be used for virtually any type of molecule or material as well as particulate carrier systems.

Thanks to the abovementioned mechanisms, under physiologic conditions, the BBB is relatively impermeable. Under pathologic conditions, a number of chemical mediators are released that increase BBB permeability. Several of these mediators of BBB opening have been studied in both in vivo and in vitro experiments and include glutamate, aspartate, taurine, ATP, endothelin-1, ATP, NO, MIP-2, tumor necrosis factor-α (TNF-α), MIP2, and IL-h, which are produced by astrocytes [36]. Other humoral agents reported to increase BBB permeability are bradykinin, 5-HT, histamine, thrombin, UTP, UMP, substance P, quinolinic acid, plateletactivating factor, and free radicals [37]. The source of these BBB-modulating mediators is of interest. Some of these agents are released by endothelium and endothelium itself responds to the released agents. For example, endothelin (ET-1) acts on ETA receptors. Under physiologic conditions, nerve terminals of neurons running close to blood vessels release mediators, such as histamine, substance P, and glutamate, which influence BBB permeability.

The BBB is poorly developed in brain tumor leading to increased vascular permeability [38]. Investigations have shown that there is opening of interendothelial TJ in human gliomas [39] and metastatic adenocarcinoma [40]. The expression of the TJ protein claudin-1 is lost in the microvessels of GBM, whereas claudin-5 and occludin are significantly downregulated and ZO-1 expression is unaffected [41]. A loss of 55 kDa occluding expression in microvessels, observed in astrocytoma and metastatic adenocarcinoma, may also contribute to endothelial TJ opening [42].

The explanation for loss of TJ molecules in brain tumor microvessels is not clear. However, VEGF, cytokines [43], and Scatter factor or hepatocyte growth factor [44] secreted by astrocytoma and other brain tumors may be involved in downregulating TJ molecules leading to TJ opening, increased vascular permeability, and cerebral edema. It is also possible that poorly differentiated neoplastic astrocytes do not release factors necessary for BBB function.

Since cerebral edema is an important consequence of brain tumor, water channel molecule, AQP4, has been examined in brain tumor by several investigators. AQP4 is massively upregulated in astrocytoma and metastatic adenocarcinoma and this correlates with BBB opening assessed by contrast-enhanced computed tomograms [45]. Mice deficient in AQP4 have a much better survival than wild-type mice in a model of brain edema caused by acute water intoxication. Upregulation of AQP4 has also been noted in rat models of ischemia [46] and brain injury [47]. Thus, it seems that breakdown of the BBB associated with brain tumors and other forms of brain injury increases the expression of AQP4. However the exact mechanism of AQP4 upregulation in different clinical situations is not known.

NANOPARTICLES

Nanoscience has a huge potential to bring benefits in areas as diverse as drug development, water decontamination, information and communication technologies, and the production of stronger, lighter materials. Nanotechnology is an emerging field that deals with interactions between cells and engineered molecules. The "nano" concept is also applied to the field of medicine, being nanomedicine used in drug delivery systems, and in cancer diagnosis and therapy. A nanoparticle is the smallest unit (10^{-9} m) that can still behave as a whole entity in terms of properties and transport and is able to cross biological membranes and access cells, tissues, and organs (Fig. 8.2) [48]. Thus, the use of nanoparticles in brain tumors treatment seems to be very promising because it has the ability to overcome the obstacle represented by the BBB. It also adds specificity and selectivity to the drugs delivered across the BBB focusing only to cancer cells, so avoiding unwilling side effects in the surrounding healthy tissues [49]. The first application of a targeted nanosystem for drug delivery concerned liposomes and it was reported in 1980 [50].

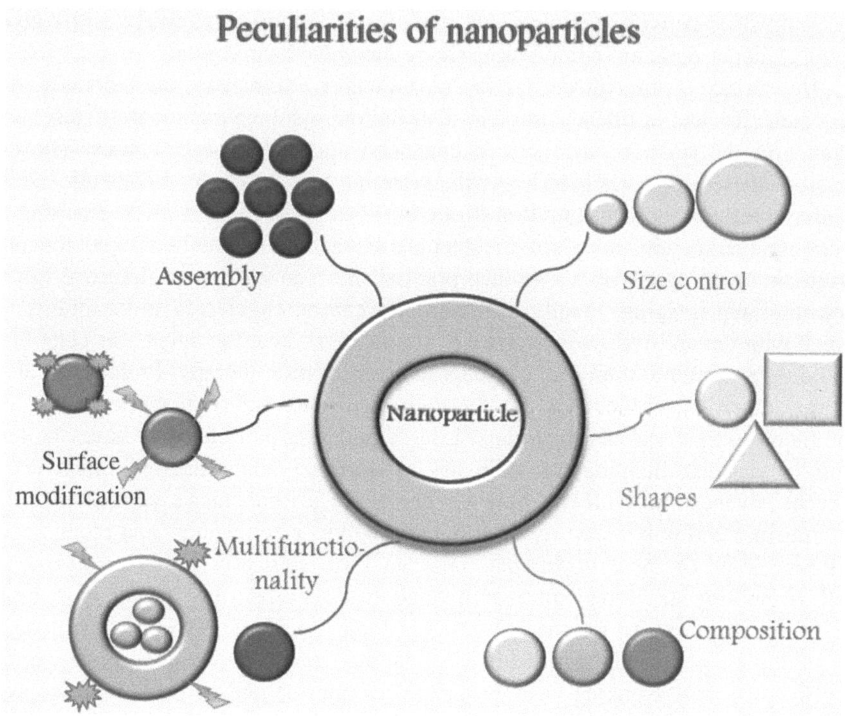

FIGURE 8.2

Nanoparticle properties.

The peculiarities of nanoparticles (NPs) (size, electrostatic charge, and lipophilic characteristics) allow them to enter the brain tissue freely, bypassing the BBB and carrying drugs. Tumor angiogenesis possesses leaky vasculature that impedes drug delivery: an abnormal basement membrane and fissures between the endothelial cells due to an absent pericytes lining. That together with a poor lymphatic drainage system causes a differential interstitial pressure at the center of tumors compared with that at the periphery. Thus, molecules ranging from approximately 10 to 100 nm accumulate in the tumor and are retained for longer, unlike the uncoated smaller drugs cleared by kidneys. This is called "enhanced permeability and retention" (EPR) effect. The retention time of drugs packed in NPs is 10× higher than that of unpacked drugs [51]. Hence the EPR effect attributed to the leaky vasculature is considered a boon for drug-delivery systems within the nanosize range.

When administered systemically, nanoparticles (NPs) can protect the loaded drugs from degradation. Small therapeutic molecular agents that are normally poorly distributed can be incorporated into NPs via a variety of chemical methods, including encapsulation, adsorption, and covalent linkage, while macromolecules can be attached to the surface of NPs to improve targeting. Reduction of toxicity to peripheral organs can also be achieved with these systems [14]. Targeted drug-delivery systems can convey drugs more effectively and conveniently, increase patient compliance, and reduce healthcare costs.

The surface of the nanocarrier can be engineered to increase the blood circulation half-life and influence the bio-distribution, while attachment of targeting ligands to the surface can result in enhanced uptake by target tissues. The net result of these properties is to lower the systemic toxicity of the therapeutic agent, while increasing the concentration of the agent in the area of interest, resulting in a higher therapeutic index for the therapeutic agent. In addition to therapeutic drugs, imaging agents can also incorporated into nanocarriers to improve tumor detection and imaging. Finally, nanoparticles can be engineered to be multifunctional with the ability to target diseased tissue, carry imaging agents for detection, and deliver multiple therapeutic agents for combination therapy. Prolonged circulation properties are ideal for slow or controlled release of therapeutic agents into the blood to treat vascular disorders. Long circulating particles may have application in vascular imaging, or even act as artificial nanoscale red blood cells. Recent advances in synthetic polymer chemistry afford precise control over the architecture and polydispersity of polymers, polymer-conjugates, and block copolymers. Some of these novel materials can form sterically stabilized nanoscale self-assembling structures with macrophage-evading properties. A unique attribute of nanoplatform-based delivery systems is their multifunctionality, characterized by multiple components, which include imaging agents, therapeutic agents, targeting ligands, and "cloaking" agents that avoid interference with the immune system. Nanotheranostic platforms are powerful tools for imaging and treatment of cancer. Multifunctionality of these systems offers a number of advantages over conventional agents. These include targeting to a diseased site thereby minimizing systemic toxicity, the ability to solubilize hydrophobic or labile drugs leading to improved pharmacokinetics and their potential to image, treat and predict therapeutic response. Targeted nanoparticle-based treatment technologies with diagnostic capabilities are referred to as theranostic agents as they form a class of agents, which can serve diagnostic and therapeutic functions simultaneously. In the current state of technology, tumor detection and therapy are performed separately. A more efficient and effective method can be achieved with theranostic nanoparticles, which would integrate the efforts for detection, treatment and follow-up monitoring of tumor response, and assist in the decision-making process for the need for further treatment.

Nanomaterial-based agents that are specific for nanomedicine include mainly polymer- or lipid-based carriers. Drugs can be absorbed onto the surface, entrapped inside, or dissolved within the matrix of these vehicles. Nanocarriers can be also combined into cells both using transfection agents, often toxic and not clinically approved, and without that agents [52]. Actual research has developed multifunctional NPs able to respond to the environment, so facilitating a more effective drug delivery. The diversity of delivery systems allows NPs to have different arrays of shapes, sizes, and components tailored for specific applications.

NPs have been shown to enter inside cells via passive transport [53] and active endocytosis [54–56]. Once inside the cells, NPs are transported to the endo-lysosomal system, where they are usually destroyed. Anyway, polylactide (PLA) and polyglycolide (PLGA)-NPs as well as lipid nanocapsules are able to escape the lysosomal compartment by disrupting the integrity of the lysosome membrane [57–59]. This endo-lysosomal escape leads to NP accumulation in the cytoplasm whose effects are unknown yet.

Among the numerous molecules studied, liposomes, gold-NP, and graphene-NP-derived seem to be among the most promising.

Liposomes are vesicles made up of a lipid bilayer, resembling a cell membrane. The lipids form a bilayer based on hydrophobic interactions in continuous parallel packing, with the hydrophilic head groups positioned toward the aqueous environment. They possess advantages of carrying hydrophilic, lipophilic, as well as amphoteric drug molecules, either entrapped inside it or on its micellar surface. Liposomes prove ideal carriers for biological agents such as siRNA because of their stable aqueous core. Moreover, it is possible to combine RNA-interfering strategies with traditional chemotherapeutics. One example is the Raf/MEK/extracellular signal-related kinase (ERK) pathway, which is essential for cellular proliferation, and found to be aberrant in several cancers [60]. As a result, several inhibitors of key proteins in the cascade have been developed as potential chemotherapeutics. Recently, it has been demonstrated that liposomes encapsulating an Mcl1-specific siRNA (siMcl1) and a chemical MEK inhibitor (PD0325901) showed a valid antitumor efficacy in vitro and in vivo [61].

Glioma cells show an upregulation of expression of IL-13 receptor $\alpha 2$ on their surface cells. In a recent study the improvement of internalization of doxorubicin-loaded nanoliposomes, targeted with conjugated IL-13, and cytotoxicity in U251 glioma cells has been shown. In an in vivo animal model the authors demonstrated the inhibition of the growth of subcutaneously implanted gliomas [62]. In anticancer gene therapy the efficiency of liposomes has been increased through surface ligand targeting, via monoclonal antibodies to specific receptors upregulated in glioma cells surface such as transferring receptors, LDL receptors, and IL-13 receptors.

For glioma gene therapy, viral vectors have been used to deliver suicide genes, pro-apoptotic genes, p53, cytokines, and caspases. These studies have shown promising preclinical results, but clinical trials have been limited by the fact that transduced cells were found only within a very short distance of the delivery site [63].

A study by Jaworski and colleagues [64] revealed that graphene platelets are toxic to U87 and U118 glioma cell lines, indicating their potential therapeutic applicability. Because of their large surface area, graphene platelets did not enter into glioma cells, but adhered to them: graphene electrons interacted with cell membranes and receptors blocking the supply of nutrients, inducing stress and activating apoptosis. Graphene is a carbon allotrope with a bidimensional hexagonal structure and with its derivatives, such as its oxide (GO), has shown potentials in numerous fields [65]. The importance of the functionalization of GO has been the key for the biological and biomedical applications of graphene which range from targeting controlled drug/gene delivery, photothermal and photodynamic cancer therapy, biological sensing and imaging, to multifunctional nanoplatforms [66]. The photothermal activity of graphene has also been investigated in the treatment of brain tumors. A recent study combined the chemo-photothermal targeted therapy of glioma within one novel multifunctional drug delivery system using a targeting peptide (IP)-modified mesoporous silicacoated graphene nanosheet (GSPI). Doxorubicin was conjugated with the GSPI-based system (GSPID), showing synergistic chemo-photothermal properties. Cytotoxicity experiments demonstrated a higher rate of death of glioma cells [67].

Tian et al. [68] showed the feasibility of encapsulating the temozolomide [(3,4-dihydro-3-methyl-4-oxoimidazo [5,1-*d*]-as-tetrazine-8-carboxamide (TMZ)], an alkylating agent, into polybutylcyanoacrylate (PBCA) polymeric nanoparticle NPs by polymerization. Based on the pattern of distribution in body organs, higher concentrations of TMZ were detected in the brain after binding to PBCA nanoparticles coated with polysorbate-80, which may be more useful for treating brain tumors.

Thermosensitive magnetic NPs were prepared with the magnetic NPs covered by a thermosensitive polymer with a critical temperature of 40–45°C. These systems were suitable for hyperthermia treatment of cancers such as brain cancer. The thermosensitive polymer shell could be ruptured at the site of action by applying an external magnetic field to increase the temperature of the inner magnetic core [69]. The NanoTherm therapy, also termed magnetic fluid hyperthermia, combined with fractionated stereotactic radiotherapy, is a new local heat treatment of solid tumors (such as glioblastoma multiforme and prostate carcinoma). Three major components are required for NanoTherm therapy: NanoTherm, Nanoplan, and NanoActivator F100 (MagForce Nanotechnologies AG, Berlin, Germany). NanoTherm is a magnetofluid consisting of superparamagnetic iron oxide nanoparticles (SPIONs), which are colloidally dispersed in water with a high iron concentration. The iron oxide magnetite (Fe_3O_4) core is approximately 12 nm in diameter and coated with an aminosilane-type shell. Due to their aminosilane coating, these small magnets can be finely dispersed in water, forming a colloidal solution that is dispensable with a syringe. Once inside the alternating magnetic field applicator, NanoActivator, these specifically designed nanoparticles are responsible for the production of warmth. Through this high-frequency magnetic field, the nanoparticles begin to oscillate and warmth is produced from directly within the tumor tissue. Depending on the temperature reached and length of treatment, the tumor cells are either directly destroyed or sensitized for the accompanying chemotherapy or radiation. NanoTherm therapy has been used successfully to treat glioblastoma multiforme in a Phase II GBM trial [70].

Nanoparticles can also be of aid in diagnosis.

Gold nanoparticles (GNPs) exhibit unique physicochemical properties, including the ability to bind amine and thiol groups, allowing surface modification and use in biomedical applications. GNPs are used to prepare nanoshells composed of gold and copper, or gold and silver to function as contrast agents in MRI, and gold–silica for photothermal ablation of tumor cells. Classically, GNPs enter into cells with a nonspecific receptor-mediated endocytosis mechanism [71]. In vivo GNPs passively accumulate at tumor sites that have leaky immature vasculature with wider fenestrations than normal mature blood vessels.

Thanks to its unique properties as contrast enhancement medium, graphene reveals accuracy in brain and spinal cord imaging. Sheets of nano-graphene are intrinsically photoluminescent, thus able to provide in vivo high-resolution live cell imaging with high quality and low cytotoxicity [72]. In addition, thanks to this property, graphene can be used also intraoperatively to obtain fluorescent imaging both during tumor and aneurysm surgery [73].

Actual molecular diagnosis avails itself of detecting the presence of specific sequences of DNA, RNA, or proteins by using biosensors. Applying engineered nanosensors, namely, nanopores, for this purpose, gives faster results than previous methods [74]. The versatility of graphene confers the capacity to detect not only single nucleotides, but also neurotransmitters, opening the possibility to its employment also for molecular diagnosis [75,76]. The affinity showed with fluorescent molecules gives space to test graphene biosensors based on fluorescence resonance energy transfer, which provide quantitative and real-time imaging of intra- and intercellular proteins' interactions, DNA, and miRNA [77–79]. In addition, it is possible to use graphene as a nanoscaffold for devices that improve the sensitivity and specificity of current biosensors.

Although promising in vitro results have been reported, it remains unclear how effective such a system would be due to intra- and interindividual patient heterogeneity considering also that toxicity of NPs in humans has not been tested yet. The majority of studies are performed in experimental settings, both in vitro and in animal models, but their translation into the clinics remains difficult due to human variability and safety-related issues.

CONCLUSION AND CRITICAL ISSUES

As summarized in the previous chapters, nanomedicine seems the efficacious answer to brain tumor treatment. It reduces hospital stays, lowers morbidity and mortality, and in a prospective view large diffusion can reduce health costs for long-term patients.

However, nanomedicine encounters difficulties in becoming widely used. It is a not completely clarified field. Toxic effects of the interactions between NPs and biological systems are not definitively assessed in humans.

There are people daily exposed to NPs originating from combustion, welding, and biomedical applications, e.g., workers in industries such as cars, aerospace, electronics, or chemical. Notwithstanding, a complete evaluation of interactions between NPs and biological systems is lacking.

It is known that NPs exhibit toxic manifestations and can result in allergy, fibrosis, organ failure, and various kind of toxicities (nephro-, hematological, neuro-, hepatic, splenic, and pulmonary) [80,81].

In addition, NPs surfaces are involved in catalytic and oxidative processes potentially cytotoxic. High dose of single-walled carbon nanotubes origins reactive oxygen species (ROS), mitochondrial dysfunction, oxidative stress, and changes in cell morphology when incubated with keratinocytes and bronchial epithelial cells. Intratracheal instillation of carbon nanotubes in rodents caused chronic lung inflammation with foreign body granuloma formation and interstitial fibrosis [82].

It is also not exactly known how long NPs remains into the body and if, during this time they can interact with healthy cells. Carbon nanotubes and quantum dots, e.g., are long-lasting remaining in the body for a long period, this make them potentially toxic and limits their use for reiterated treatments. Copper, iron, gold, iron, manganese, titanium, silica, and other carbon-based nanomaterials are some of the NPs to which humans are exposed significantly and may cause several health-related problems including neurotoxicity [83]. The effect of NPs on the cell membrane may be due to their direct toxicity, or indirectly, they may induce some cascade mechanism that disrupts the tight junctions in the BBB or alters the permeability of the membrane. It has been shown that intravenous, intraperitoneal, or intracerebral administration of Ag, Cu, or Al NPs disrupts the BBB, as indicated by staining with albumin-bound Evans blue [84].

Another example is nanoparticulate TiO_2, used to develop cosmetics, foodstuffs, toothpaste, sun blocks, printing ink, car materials, cleaning products, materials for industrial photocatalytic applications including solar cells, and catalysts for remediation of organic matter in wastewater. TiO2 NPs accumulate in the brain and induce structural changes in the neuronal structure [85,86]. Altered gene expression was detected for prenatal TiO_2 NP exposure, which was involved in cell death, brain development, and response to oxidative stress in newborn pups [87].

Iron oxide or SPIONs are small NPs composed of a Fe_3O_4 (magnetite) or Fe_2O_3 (maghemite) core. Their potential application ranges from biomedical imaging (magnetic resonance imaging, positron emission tomography, or ultrasound as contrast agent), gene and drug delivery, tissue regeneration, hyperthermia in cancer treatment, catalysis, and magnetic storage. They are extensively used specifically for brain imaging or brain-targeted drug and gene delivery, due to their ability to move across the BBB [88]. SPIONs shown a potential toxicity that can lead to altered gene expression, actin modulation, interference with cell cycle regulation and signaling pathways, excessive ROS generation, and disruption of iron homeostasis. Fe_3O_4 NPs also had a substantial cytotoxic effect on PC12 cells by modulating the cell cycle and inducing apoptosis [89].

Concerning carbon-based nanomaterials, they have a potential use in a variety of biomedical applications, including early diagnosis of cancer, imaging, targeted photothermal therapy, drug delivery, and tissue engineering. Studies of carbon nanomaterials have indicated the potential neurotoxic effects after inhalation or systemic exposure. A study has shown that oxidative stress is involved in this toxic pathway, with surface coating playing an important role [90]. In vivo studies demonstrate systemic bio-distribution and biopersistence of graphene following intravenous delivery. Similar to other foreign bodies, graphene has the potential to induce foreign body tumors, thus long-term adverse health impacts has to be considered. The limited literature on in vitro toxicity suggests that graphene can be either benign or toxic to cells, its biological response depending on layers number, lateral size, stiffness, hydrophobicity, surface functionalization, and dose. Direct or indirect generation of ROS unbalanced by the cellular antioxidant enzymes is currently the main mechanism proposed for graphene toxicity [91]. It has been observed that GO directly interacts with the cell membranes leading to their physical damage [92]. In vitro studies agree that GO promotes cytotoxicity mainly generating ROS in a dose-dependent manner [93]. Graphene is also genotoxic, penetrating into the cells and causing damage of DNA such as fragmentation and/or chromosomal aberrations even for low doses at short time of exposure [94].

Notwithstanding, there are others potential risks related with this novel approach. Some cancer cell types could develop drug resistance making ineffective the drugs released from the targeted NPs. Moreover, NPs might change stability, solubility, and pharmacokinetic properties of the carried drugs. In addition, some materials used to create NPs possess low toxicity but degrade quickly and do not circulate in tissues long enough to warrant a sustained drug/gene delivery.

Objects of debate are the results about the long-term effects of interactions between NPs and coating of molecules and target cells. In order for this promising field to rapidly progress, focus must be placed on elucidating the safety of these novel materials.

REFERENCES

[1] Louis DN, Ohgaki H, Wiestler OD, Cavenee WK, Burger PC, Jouvet A, et al. The 2007 WHO classification of tumours of the central nervous system. Acta Neuropathol 2007;114(2):97–109.

[2] Scott JN, Rewcastle NB, Brasher PMA, et al. Which glioblastoma multiforme patient will become a long-term survivor? A population-based study. Ann Neurol 1999;46:183–8.

[3] McLendon RE, Halperin EC. Is the long-term survival of patients with intracranial glioblasoma multiforme overstated? Cancer 2003;98(8):1745–8.

[4] Hottinger AF, Yoon H, DeAngelis LM, Abrey LE. Neurological outcome of long-term glioblastomas survivors. J Neurooncol 2009;95(3):301–5.

[5] Stupp R, Mason WP, van den Bent MJ, et al. Radiotherapy plus concomitant and adjuvant temozolomide for glioblastoma. N Engl J Med 2005;352(22):987–99.

[6] Stupp R, Hegi ME, Mason WP, et al. Effects of radiotherapy with concomitant and adjuvant temozolomide versus radiotherapy alone on survival in glioblastoma in a randomized phase III study: 5-year analysis of the EORTC-NCIC trial. Lancet Oncol 2009;10(5):459–66.

[7] Johnson DR, Ma DJ, Buckner JC, Hammack JE. Conditional probability of long-term survival in glioblastoma. Cancer 2012;118(22):5609–13.

[8] Lonser RR, et al. Induction of glioblastoma multiforme in nonhuman primates after therapeutic doses of fractionated whole-brain radiation therapy. J Neurosurg 2002;97(6):1378–89.

[9] Dehdashti AR, Sharma S, Laperriere N, Bernstein M. Coincidence vs cause: cure in three glioblastoma patients treated with brachytherapy. Can J Neurol Sci 2007;34:339–42.

[10] Haratmann C, Hentschel B, Simon M, et al. Long-term survival in primary glioblastoma with versus without isocitrate dehydrogenase mutations. Clin Cancer Res 2013;19(18):5146.

[11] Westphal M, Lamszus K. The neurobiology of gliomas: from cell biology to the development of therapeutic approaches. Nat Rev Neurosci 2011;12(9):495–508.

[12] Louis DN, Perry A, Reifenberger G, von Deimling A, Figarella-Branger D, Cavenee WK, et al. The 2016 World Health Organization classification of tumors of the central nervous system: a summary. Acta Neuropathol 2016;131(6):803–20.

[13] Ohgaki H, Kleihues P. The definition of primary and secondary glioblastoma. Clin Cancer Res 2013;19(4):764–72.

[14] Caruso G, Caffo M, Alafaci C, Raudino G, Cafarella D, et al. Could nanoparticles systems have a role in the treatment of cerebral gliomas? Nanomedicine 2011;7(6):744–52.

[15] Caruso G, Raudino G, Caffo M, Alafaci C, Granata F, et al. Nanotechnology platforms in diagnosis and treatments of primary brain tumors. Recent Pat Nanotechnol 2010;4(2):119–22.

[16] Stupp R, Hegi ME, van den Bent MJ, Mason WP, Weller M, Mirimanoff RO, et al. Changing paradigms: an update on the multidisciplinary management of malignant glioma. Oncologist 2006;11(2):165–80.

[17] Ruben JD, Dally M, Bailey M, Smith R, McLean CA, Fedele P. Cerebral radiation necrosis: Incidence, outcomes, and risk factors with emphasis on radiation parameters and chemotherapy. Int J Radiat Oncol Biol Phys 2006;65(2):499–508.

[18] Caffo M, Barresi V, Caruso G, Cutugno M, La Fata G, Venza M, et al. Innovative therapeutic strategies in the treatment of brain metastases. Int J Mol Sci 2013;14(1):2135–74.

[19] Sawyer AJ, Piepmeier JM, Saltzman WM. New methods for direct delivery of chemotherapy for treating brain tumors. Yale J Biol Med 2006;79(3–4):141–52.

[20] Caruso G, Caffo M, Raudino G, Alafaci C, Tomasello F. Nanomedicine and brain tumors treatment. In: Souto EB, editor. Patenting nanomedicines: Legal aspects, intellectual property and grant opportunities. Berlin Heidelberg: Springer-Verlag; 2012. p. 167–204.

[21] Mamelak AN. Locoregional therapies for glioma. Oncology 2005;19(14):1803–10.

[22] Sawyer AJ, Saucier-Sawyer JK, Booth CJ, Liu J, Patel T, Piepmeier JM, et al. Convection-enhanced delivery of camptothecin-loaded polymer nanoparticles for treatment of intracranial tumors. Drug Deliv Transl Res 2011;1(1):34–42.

[23] Hu CJ, Zhang L. Therapeutic nanoparticles to combat cancer drug resistance. Curr Drug Metab 2009;10(8):836–41.

[24] Latif AZ, Signorini D, Gregor A, Whittle IR. The costs of managing patients with malignant glioma at a neuro-oncology clinic. Br J Neurosurg 1998;12(2):118–22.

[25] Goldberg MS, Hook SS, Wang AZ, Bulte JW, Patri AK, Uckun FM, et al. Biotargeted nanomedicines for cancer: six tenets before you begin. Nanomedicine (Lond) 2013;8(2):299–308.

[26] Park S, Ruoff RS. Chemical methods for the production of graphenes. Nat Nanotechnol 2009;4(4):217–24.

[27] Mattei TA, Rehman AA. Technological developments and future perspectives on graphene-based metamaterials: a primer for neurosurgeons. Neurosurgery 2014;74 (5):499–516.

[28] Egleton RD, Davis TP. Bioavailability and transport of peptides and peptide drugs into the brain. Peptides 1997;18:1431–9.

[29] Cardoso FL, Brites D, Brito MA. Looking at the blood–brain barrier: molecular anatomy and possible investigation approaches. Brain Res Rev 2010;64:328–63.

[30] Hawkins BT, Davis TP. The blood–brain barrier/neurovascular unit in health and disease. Pharmacol Rev 2005;57:173–85.

[31] Urquhart BL, Kim RB. Blood–brain barrier transporters and response to CNS-active drugs. Eur J Clin Pharmacol 2009;65:1063–70.

[32] Kemper EM, Boogerd W, Thuis I, Beijnen JH, van Tellingen O. Modulation of the blood–brain barrier in oncology: therapeutic opportunities for the treatment of brain tumours? Cancer Treat Rev 2004;30:415–23.

[33] Pardridge WM. Blood–brain barrier delivery. Drug Discov Today 2007;12:54–61.

[34] Moos T, Morgan EH. The metabolism of neuronal iron and its pathogenic role in neurological disease: review. Ann NY Acad Sci 2004;1012:14–26.

[35] Jain S, Mishra V, Singh P, Dubey PK, Saraf DK, Vyas SP. RGD-anchored magnetic liposomes for monocytes/neutrophils-mediated brain targeting. Int J Pharm 2003;261:43–55.

[36] Abbott NJ. Inflammatory mediators and modulation of blood–brain barrier permeability. Cell Mol Neurobiol 2000;20:131–47.

[37] Abbott NJ. Astrocyte – endothelial interactions and blood–brain barrier permeability. J Anat 2002;200:629–38.

[38] Groothuis DR, Vriesendorp FJ, Kupfer B, Warnke PC, Lapin GD, Kuruvilla A, et al. Quantitative measurements of capillary transport in human brain tumors by computed tomography. Ann Neurol 1991;30:581–8.

[39] Long DM. Capillary ultrastructure and the blood–brain barrier in human malignant brain tumors. J Neurosurg 1970;32:127–44.
[40] Long DM. Capillary ultrastructure in human metastatic brain tumors. J Neurosurg 1979;51:53–8.
[41] Liebner S, Fischmann A, Rascher G, Duffner F, Grote EH, Kalbacher H, et al. Claudin-1 and claudin-5 expression and tight junction morphology are altered in blood vessels of human glioblastoma multiforme. Acta Neuropathol (Berl) 2000;100:323–31.
[42] Papadopoulos MC, Saadoun S, Woodrow CJ, Davies DC, Costa-Martins P, Moss RF, et al. Occludin expression in microvessels of neoplastic and non-neoplastic human brain. Neuropathol Appl Neurobiol 2001;27:384–95.
[43] de Vries HE, Blom-Roosemalen MC, van Oosten M, de Boer AG, van Berkel TJ, Breimer DD, et al. The influence of cytokines on the integrity of the blood–brain barrier in vitro. J Neuroimmunol 1996;64:37–43.
[44] Lamszus K, Laterra J, Westphal M, Rosen EM. Scatter factor/hepatocyte growth factor (SF/HGF) content and function in human gliomas. Int J Dev Neurosci 1999;17:517–30.
[45] Saadoun S, Papadopoulos MC, Davies DC, Krishna S, Bell BA. Aquaporin-4 expression is increased in oedematous human brain tumours. J Neurol Neurosurg Psychiatry 2002;72:262–5.
[46] Taniguchi M, Yamashita T, Kumura E, Tamatani M, Kobayashi A, Yokawa T, et al. Induction of aquaporin-4 water channel mRNA after focal cerebral ischemia in rat. Brain Res Mol Brain Res 2000;78:131–7.
[47] Vizuete ML, Venero JL, Vargas C, Ilundain AA, Echevarria M, Machado A, et al. Differential upregulation of aquaporin-4 mRNA expression in reactive astrocytes after brain injury: potential role in brain edema. Neurobiol Dis 1999;6:245–58.
[48] Sahoo SK, Parveen S, Panda JJ. The present and future of nanotechnology in human health care. Nanomedicine 2007;3:20–31.
[49] Caruso G, Caffo M, Merlo L, La Fata G, Passalacqua M, Tomasello F. Nanomedicine in brain tumors. In: Tiwari A, Tiwari AS, editors. Bioengineered nanomaterials. London: CRC Press; 2013. p. 295–326.
[50] Heath TD, Fraley RT, Papahdjopoulos D. Antibody targeting of liposomes: cell specificity obtained by conjugation of F(ab')2 to vesicle surface. Science 1980;210:539–41.
[51] Byrne JD, Betancourt T, Brannon-Peppas L. Active targeting schemes for nanoparticle systems in cancer therapeutics. Adv Drug Deliv Rev 2008;60:1615–26.
[52] Mailander V, Lorenz MR, Holzapfel V, Musyanovych A, et al. Carboxylated superparamagnetic iron oxide particles label cells intracellularly without transfection agents. Mol Imaging Biol 2008;10:138–46.
[53] Banerji SK, Hayes MA. Examination of nonendocytotic bulk transport of nanoparticles across phospholipid membranes. Langmuir 2007;23:3305–13.
[54] Lorenz MR, Holzapfel V, Musyanovych A, Nothelfer K, Walther P, Frank H, et al. Uptake of functionalized, fluorescent-labeled polymeric particles in different cell lines and stem cells. Biomaterials 2006;27:2820–8.
[55] Rejman J, Oberle V, Zuhorn IS, Hoekstra D. Size-dependent internalization of particles via the pathways of clathrin- and caveolae-mediated endocytosis. Biochem J 2004;377(Pt 1):159–69.
[56] Schulze Jr E, Ferrucci JT, Poss K, Lapointe L, Bogdanova A, Weissleder R. Cellular uptake and trafficking of a prototypical magnetic iron oxide label in vitro. Invest Radiol 2003;30:604–10.

[57] Panyam J, Labhasetwar V. Dynamics of endocytosis and exocytosis of poly(D,L-lactide-*co*-glycolide) nanoparticles in vascular smooth muscle cells. Pharm Res 2003;20:212–20.

[58] Panyam J, Zhou WZ, Prabha S, Sahoo SK, Labhasetwar V. Rapid endo-lysosomal escape of poly(DL-lactide-*co*-glycolide) nanoparticles: implications for drug and gene delivery. FASEB J 2002;16:1217–26.

[59] Paillard A, Hindre F, Vignes-Colombeix C, Benoit JP, Garcion E. The importance of endo-lysosomal escape with lipid nanocapsules for drug subcellular bioavailability. Biomaterials 2010;31:7542–54.

[60] Ciuffreda L, Del Bufalo D, Desideri M, Di Sanza C, Stoppacciaro A, Ricciardi MR, et al. Growth-inhibitory and antiangiogenic activity of the MEK inhibitor PD0325901 in malignant melanoma with or without BRAF mutations. Neoplasia 2009;11:720–31.

[61] Kang SH, Cho HJ, Shim G, Lee S, Kim SH, Choi HG, et al. Cationic liposomal co-delivery of small interfering RNA and a MEK inhibitor for enhanced anticancer efficacy. Pharm Res 2011;28:3069–78.

[62] Madhankumar AB, Slage-Webb B, Mintz A, Sheehan JM, Connor JR. Interleukin-13 receptor-targeted nanovesicles are a potential therapy for glioblastoma multiforme. Mol Cancer Ther 2006;5:3162–9.

[63] Sun C, Veiseh O, Gunn J, Fang C, Hansen S, Lee D, et al. In vivo MRI detection of gliomas by chlorotoxin-conjugated superparamagnetic nanoprobes. Small 2008;4:372–9.

[64] Jaworski S, Sawosz E, Grodzik M, et al. In vitro evaluation of the effects of graphene platelets on glioblastoma multiforme cells. Int J Nanomed 2013;8:413–20.

[65] Feng LZ, Liu Z. Graphene in biomedicine: opportunities and challenges. Nanomedicine 2011;6:317–24.

[66] Caffo M, Merlo L, Marino D, Caruso G. Graphene in neurosurgery: the beginning of a new era. Nanomedicine 2015;10:615–25.

[67] Wang Y, Wang K, Zhao J, Liu X, et al. Multifunctional mesoporous silica-coated graphene nanosheet used for chemo-photothermal synergistic targeted therapy of glioma. J Am Chem Soc 2013;135:4799–804.

[68] Tian X, Lin X, Wei F, Feng W, Huang Z, et al. Enhanced brain targeting of temozolomide in polysorbate-80 coated polybutylcyanoacrylate nanoparticles. Int J Nanomed 2011;6:445–52.

[69] Chang WH, Kao CH, Lin CI, Wang SJ, US0154397 (2007).

[70] Maier-Hauff K, Ulrich F, Nestler D, Niehoff H, Wust P, Thiesen B, et al. Efficacy and safety of intratumoral thermotherapy using magnetic iron-oxide nanoparticles combined with external beam radiotherapy on patients with recurrent glioblastoma multiforme. J Neurooncol 2011;103(2):317–24.

[71] Chithrani BD, Ghazani AA, Chan WC. Determining the size and shape dependence of gold nanoparticle uptake into mammalian cells. Nano Lett 2006;6:662–8.

[72] Qian J, Wang D, Cai FH, et al. Observation of multiphoton-induced fluorescence from graphene oxide nanoparticles and applications in in vivo functional bioimaging. Angew Chem Int Ed Engl 2012;51(42):10570–5.

[73] Lane BC, Cohen-Gadol AA. Fluorescein fluorescence use in the management of intracranial neoplastic and vascular lesions: a review and report of a new technique. Curr Drug Discov Technol 2013;10(2):160–9.

[74] Stine R, Robinson JT, Sheehan PE, Tamanaha CR. Real-time DNA detection using reduced graphene oxide field effect transistors. Adv Mater 2010;22(46):5297–300.

[75] Bonanni A, Pumera M. Graphene platform for hairpin-DNA-based impedimetric genosensing. ACS Nano 2011;5(3):2356–61.
[76] Wang Y, Li Y, Tang L, Lu J, Li J. Application of graphene-modified electrode for selective detection of dopamine. Electrochem Commun 2009;11(4):889–92.
[77] Kasry A, Ardakani AA, Tulevski GS, Menges B, Copel M. Vyklicky, highly efficient fluorescence quenching with graphene. J Phys Chem 2012;116(4):2858–62.
[78] He S, Song B, Li D, et al. A graphene nanoprobe for rapid, sensitive, and multicolor fluorescent DNA analysis. Adv Funct Mater 2010;20(3):453–9.
[79] Lu Z, Zhang L, Deng Y, Li S, He N. Graphene oxide for rapid microRNA detection. Nanoscale 2012;4(19):5840–2.
[80] Nel A, Xia T, Madler L, Li N. Toxic potential of materials at the nanolevel. Science 2006;311(5761):622–7.
[81] Singh N, Manshian B, Jenkins GJS, et al. NanoGenotoxicology: the DNA damaging potential of engineered nanomaterials. Biomaterials 2009;30(23–24):3891–914.
[82] Lesniak MS, Upadhyay U, Goodwin R, Tyler B, Brem H. Local delivery of doxorubicin for the treatment of malignant brain tumors in rats. Anticancer Res 2005;25:3825–31.
[83] Caffo M, Caruso G, La Fata G, Barresi V, et al. Heavy metals and epigenetic alterations in brain tumors. Curr Genom 2014;15:457–63.
[84] Ahmed R, Oborski MJ, Hwang M, Lieberman FS, Mountz JM. Malignant gliomas: current perspectives in diagnosis, treatment, and early response assessment using advanced quantitative imaging methods. Cancer Manag Res 2014;6:149–70.
[85] Wang J, Chen C, Liu Y, et al. Potential neurological lesion after nasal instillation of TiO_2 nanoparticles in the anatase and rutile crystal phases. Toxicol Lett 2008;183:72–80.
[86] Wang J, Liu Y, Jiao F, et al. Time-dependent translocation and potential impairment on central nervous system by intranasally instilled TiO_2 nanoparticles. Toxicology 2008;254:82–90.
[87] Shimizu M, Tainaka H, Oba T, et al. Maternal exposure to nanoparticulate titanium dioxide during the prenatal period alters gene expression related to brain development in the mouse. Part Fibre Toxicol 2009;6:20.
[88] Kong SD, Lee J, Ramachandran S, et al. Magnetic targeting of nanoparticles across the intact blood-brain barrier. J Control Release 2012;164:49e57.
[89] Wang B, Feng W, Zhu M, et al. Neurotoxicity of low-dose repeatedly intranasal instillation of nano- and submicronsized ferric oxide particles in mice. J Nanopart Res 2009;11:41e53.
[90] Zhang Y, Xu Y, Li Z, et al. Mechanistic toxicity evaluation of uncoated and PEGylated single-walled carbon nanotubes in neuronal PC12 cells. ACS Nano 2011;5:7020–33.
[91] Sanchez VC, Jachak A, Hurt RH, Kane AB. Biological interactions of graphene-family nanomaterials: an interdisciplinary review. Chem Res Toxicol 2012;25(1):15–34.
[92] Hu W, Peng C, Lv M, et al. Protein corona-mediated mitigation of cytotoxicity of graphene oxide. ACS Nano 2011;5(5):3693–700.
[93] Gollavelli G, Ling YC. Multi-functional graphene as an in vitro and in vivo imaging probe. Biomaterials 2012;33(8):2532–45.
[94] Akhavan O, Ghaderi E, Emamym H, Akhavan F. Genotoxicity of graphene nanoribbons in human mesenchymal stem cells. Carbon 2013;54:419–31.

Index

Note: Page numbers followed by "*f*" and "*t*" refer to figures and tables, respectively.

O

P

Q

R

FSC
www.fsc.org
MIX
Papier | Fördert gute Waldnutzung
FSC® C083411

Kolibri 360 GmbH
Ferdinand-Jühlke-Straße 7
99095 Erfurt, Deutschland
produktsicherheit@kolibri360.de